PHalarope Books

PHalarope Books are designed specifically for the amateur naturalist. These volumes represent excellence in natural history publishing. Most books in the PHalarope series are based on a nature course or program at the college or adult education level or are sponsored by a museum or nature center. Each PHalarope book reflects the author's teaching ability as well as writing ability. Among the books:

Botany in the Field: An Introduction to Plant Communities for the Amateur Naturalist
Jane Scott

The Curious Naturalist
John Mitchell and the Massachusetts Audubon Society

A Field Guide to the Familiar: Learning to Observe the Natural World
Gale Lawrence

Insect Life: A Field Entomology Manual for the Amateur Naturalist
Ross H. Arnett, Jr., and Richard L. Jacques, Jr.

A Natural History Notebook of North American Animals
National Museum of Natural History, Canada

Nature with Children of All Ages: Activities and Adventures for Exploring, Learning, and Enjoying the World Around Us
Edith A. Sisson, the Massachusetts Audubon Society

Owls: An Introduction for the Amateur Naturalist
Gordon Dee Alcorn

The Plant Observer's Guidebook: A Field Botany Manual for the Amateur Naturalist
Charles E. Roth

Pond and Brook: A Guide to Nature Study in Freshwater Environments
Michael J. Caduto

The Seaside Naturalist: A Guide to Nature Study at the Seashore
Deborah A. Coulombe

The Sky Observer's Guidebook
Charles E. Roth

Suburban Geology: An Introduction to the Common Rocks and Minerals of Your Back Yard and Local Park
Richard Headstrom

Suburban Wildflowers: An Introduction to the Common Wildflowers of Your Back Yard and Local Park
Richard Headstrom

Suburban Wildlife: An Introduction to the Common Animals of Your Back Yard and Local Park
Richard Headstrom

Thoreau's Method: A Handbook for Nature Study
David Pepi

365 Starry Nights: An Introduction to Astronomy for Every Night of the Year
Chet Raymo

Trees: An Introduction to Trees and Forest Ecology for the Amateur Naturalist
Laurence C. Walker

The Wildlife Observer's Guidebook
Charles E. Roth, Massachusetts Audubon Society

Wood Notes: A Companion and Guide for Birdwatchers
Richard H. Wood

A NATURALIST INDOORS

Observing the World
of Nature Inside
Your Home

GALE LAWRENCE

Illustrated by Adelaide Tyrol

PHalarope
Books

A Naturalist Indoors:
Observing the World of Nature Inside Your Home
All Rights Reserved © 1986, 2000 by Gale Lawrence

AN AUTHORS GUILD BACKINPRINT.COM EDITION

Published by iUniverse.com, Inc.

For information address:
iUniverse.com, Inc.
5220 S 16th, Ste. 200
Lincoln, NE 68512
www.iuniverse.com

Originally published by Prentice Hall Press

Credit for Graphic: Adelaide Tyrol

ISBN: 0-595-16755-1

Printed in the United States of America

Dedicated to my parents, siblings, roommates, and housemates, all of whom have had to tolerate my eccentric housekeeping habits—and to my cat, Hussy, who inspired me to write this book.

ACKNOWLEDGMENTS

Many people helped me write this book.

First, I'd like to thank those friends and neighbors who donated specimens, shared experiences, or, in general, sustained an interest in my project: Lisa Barrett, Hope Bertelson, Alfred DePew, Lynn Erb, Paula Kelley, Neil Stout, and Sue von Baeyer.

I also made some new friends—and deepened some old friendships—in my obsessive pursuit of accurate information. I thank Margaret Barker of the Vermont Institute of Natural Science, Theresa Carroll of Mother Nature's Helper, Karen Claxton formerly of the Pet Menagerie, Nancy Crane of the Bailey/Howe Reference Department at the University of Vermont, Stephanie Coyne DeGhett of the English Department at SUNY Potsdam, Victor J. DeGhett of the Biology and Psychology Departments at SUNY Potsdam, Job Michael Evans of Patience of Job Training for Dogs and People, Peter Hope of Mother Nature's Helper, Bette Lamore of the Medical Microbiology Department at the University of Vermont, William Murray of Gills and Gravel Tropical Fish, Vernon Peppard of Geomap Company, Craig Robertson of the Bailey/Howe Reference Department, Don and Libby Santaw of The Garden Patch, and Ann Turner of the Pet Menagerie for answering my questions and reading parts of the manuscript.

Furthermore, I talked on the telephone or corresponded with numerous individuals and organizations who provided me with facts, opinions, and, in some instances, whole packets of useful information. My thanks to Bob Behnme of *Pet Business*; R. Winston Bell of George C. Brown and Company, Inc.; R. C. Benson of Aloe Laboratories of Texas, Inc., and Hilltop Gardens; Thomas E. Cole of the Rubber Manufacturers Association; Robin A. Critzer of the National Aloe Science Council; Marty Duffy of the Aromatic Red Cedar Closet Lining Manufacturers Association; Patricia A. Flood of the American Brush Manufacturers Association; Charles Glass of *Cactus and Succulent Journal*; Steve Hastings of *The Pet Dealer*; Richard Haubsich of the American Bamboo Society; Ray Henry of Aloe Laboratories of Texas, Inc.; George Long of the American Cavy Breeders Association; Manton Cork Corporation; Tovah Martin of Logee's Greenhouses; Bill Parsonson of the Goldfish Society of America; Neal Pronek of T. F. H. Publications, Inc.; Dorothy Roberson of the National Cotton Council of America; the Sponge and Chamois Institute; Curtis Strong of American Distilling; Roger Swain of *Horticulture*; Anne Tinari of the African Violet Society of America; and Russell H. Williams of Eberhard Faber, Inc.

I am also indebted to the many scientists who have conducted laboratory and field studies of the plants, animals, and other life forms I've discussed in my book. I am especially grateful to Richard D. Alexander for his work on crickets, Paul K. Anderson and F. H. Bronson for their work on house mice, Richard B. Fischer for his work on chimney swifts, and Daniel H. Janzen for his work on the ecology of microorganisms.

Finally, I owe special thanks to Adelaide Tyrol for illustrating my book with her fine drawings, Peter Wagner for reading early drafts of my manuscript with intelligence and sensitivity, Deborah Bouchard for typing the manuscript, and Bob Spear, who shook his head and wondered, but allowed me the space to be a naturalist indoors.

CONTENTS

III. OTHER KINGDOMS IN THE KITCHEN

IV. HOUSEHOLD ECOLOGY

V. HOUSEHOLD NATURAL HISTORY

INTRODUCTION

When I first discovered the natural world—belatedly, in my thirties—I suddenly wanted to spend all my time outdoors. I took up backpacking and spent every weekend hiking and camping. I moved from an efficiency apartment in downtown Washington, D.C., to a gatehouse on a country estate in suburban Maryland. A few years later I bought an old farmhouse on a dirt road in Vermont, where I still live. In Vermont, I equipped myself with binoculars, a hand lens, and over twenty field guides to help me identify every plant and animal I encountered. After a few more years I quit my job—I had been teaching English for twelve years—and began spending all my time learning and writing about the natural world.

Three books later, it occurred to me that I was ready to come back indoors. Having trained myself outdoors, I was ready to look at the indoor environment with a naturalist's eye. I wanted to see what it could teach me about myself and the species I belong to.

The year I spent focusing on the indoors was a confusing and contradictory time. On the one hand, I was fascinated by each plant and animal I researched, but on the other, I found myself becoming increasingly uncomfortable with my own species. Why do we need to control and manipulate—even obliterate—the creatures who live closest to us?

When I finished with pets and houseplants, I proceeded to the inhabitants of my kitchen and the wild creatures who seek food, living space, or some other resource in other parts of my home. I was beset by such complex considerations as direct competition (Am I allowed to protect my food?), sanitation (Am I allowed to protect my health?), and predator-prey relationships (Who am I to decide who should live and who should die?).

As I explored my interactions with the various organisms who live around my house, I learned that I am a "situational survivalist." Basically I prefer to live and let live, but if a specific organism threatens me in a specific way—like mildew trying to decompose my shower curtain, clothes moths trying to eat my woolens, or a germ-laden housefly trying to land on food I plan to eat—I feel justified in defending my interests against that specific organism.

Having decided that I, too, have rights in this complicated business of living indoors, I had only my guilt at the way my species exploits the natural world to contend with. I was pleasantly surprised to discover that the natural products I have surrounded myself with are all renewable resources. The natural world does indeed produce enough to offer the human species some comfort and support, but only if we harvest wisely and with respect for life processes.

The "indoors" I have considered is not just my own home. It includes the suburban houses I grew up in, my grandparents' farm, the homes of my friends, and whatever other indoor environments happened to present me with a plant or animal to think about.

Indoor environments turn out to be every bit as interesting as outdoor environments—with the difference that they reflect the tastes, values, attitudes, and housekeeping habits of the human beings who create and maintain them. I now see myself as part of two ecologies: one outdoors, where I play at best a bit part, always feeling guilty about what my species is doing and wondering where I fit in; and one indoors, where my own choices and actions actually shape the environment. I don't flatter myself that I'm in charge of my own home, but writing this book has made me a full, conscious, and increasingly tolerant participant in the ecology that operates indoors.

I. AN INDOOR
ANIMAL KINGDOM

CATS

My cat, who died at an advanced age, inspired this book. There was nothing exceptional about her. She was just a plain old house cat with an undistinguished ancestry that could probably be traced back about four thousand years to ancient Egypt, where African wild cats first entered human households as rodent controls. But I often found myself just looking at her. Her obvious satisfaction with life indoors made me mindful of the indoor environment—and all the plants and animals that we ourselves bring indoors or that find their own way in because the indoors offers them something they want. My indoor observations have led me considerably beyond my cat, but she still offers a good place to begin because she gave me thirteen years' worth of herself to think about.

The first thing she taught me was her preference for *nocturnal*, or nighttime, activity. I named her "Hussy," in fact, because she liked to stay out all night and sleep indoors most of the day. Her name began as a joke, but the behavior that inspired it taught me to respect her basic ways, She not only preferred the night—she was physically adapted to it.

Her eyes, for instance, "glowed" in the dark, as do the eyes of all nocturnal animals when they are caught in a car's headlights. These animals have a special reflecting layer behind the retina that sends whatever dim light is available through the retina a second time. When a car's headlights strike such eyes, some of the light is reflected back at the driver as eyeshine. Hussy also had sensitive, cup-shaped ears that she could flick in different directions to hear what she couldn't see. Her long facial whiskers enabled her to sense physical objects, so she could move through darkness without bumping into them.

Eventually—and with much moral confusion on my part—Hussy also taught me that she was a predator. I fed her regularly and well, including even vitamin supplements during one obsessive period, but still she presented me with so many frogs, birds, chipmunks, and flying squirrels that I began keeping her indoors to protect the local wildlife. But keeping her indoors didn't change her nature. She merely concentrated on the mice who were trying to create niches for themselves in my kitchen.

Sometimes, as Hussy and I sat quietly in the kitchen after supper, she would suddenly snap her ears forward and direct her complete attention to a cupboard door. While I hadn't heard a thing, she had detected the faint rustle or ultrasonic squeak of a mouse. What she did next—if I could bear to watch—was show me the discrete stages of a feline hunt. A cat stalks, catches, kills, and eats in four separately motivated patterns of behavior that don't always have to do with hunger.

Just hearing a sound that might be a mouse was enough to motivate Hussy's stalking behavior. She would crouch low and move across the floor almost on her belly. With her retractable claws drawn in, she moved silently on her soft footpads. When she got to within pouncing distance of the cupboard door, she would stop and poise herself for the catch, remaining absolutely motionless except for her tail, which would twitch nervously from side to side.

Hussy was a determined huntress, even if she had been fed just a few hours before. Once the mouse had engaged her stalking behavior, she seemed bent upon catching it. She would wait patiently for the mouse to emerge from the cupboard and then pounce on it with all the speed, energy, and precision stored in her tense, well-coordinated body. The one time I watched the whole horrifying business from beginning to end, I saw my sweet house cat land on the mouse with her front claws fully spread, grab it in her sharp teeth, shake it viciously, and bat it around as if a much bigger animal were fighting back. Then she bit it squarely at the base of the neck. Having witnessed the stalking, catching, and killing stages of the feline hunt right there in my kitchen, I was grateful that my resident predator preferred to eat in the privacy of another room.

Hussy also taught me—and needed to remind me on a regular basis—that a cat is a solitary, independent animal not always in the mood for human attentions. Sometimes she would reject my social gestures completely, walking away from me when I approached her or jumping off my lap as soon as I picked her up. Other times she would just look at me with a steady, sober stare that kept me at a distance.

When I began thinking about why a solitary, nocturnal animal would want to live indoors with a gregarious, day-active human being—or why a human being would want such an animal indoors—I had to look to other components of my relationship with Hussy. She often behaved like a kitten, even after the veterinarian had declared her a geriatric cat. She had a rich and kittenish fantasy life, which consisted of hiding in bags or boxes and sneaking up on things that were standing still. She loved to attack the wadded pieces of paper I had ripped out of my typewriter and thrown on my study floor. And sometimes she liked to purr in my lap as I read or curl up next to me when she slept.

This kittenish behavior is what distinguishes a domestic cat from a wild cat. Theorists of the cat's domestication point out that wild cats, despite their essential solitude, begin life as social creatures and become solitary only as adults. The domestic cat's ability to socialize with human beings derives from the kitten's repertoire of social behaviors, which may accidentally have persisted into adulthood in the ancestors of today's domestic cats.

Interestingly enough, human beings themselves seem necessary to the expression of this residual kittenishness. If a domestic kitten is born outdoors and grows up without human contact, it develops into an adult as antisocial as any wild cat. Even if a thoroughly socialized adult house cat is abandoned by its human companions, it will eventually abandon its social behavior and soon become the solitary predator a cat is designed to be.

Hussy showed me both sides of her nature and gradually taught me what kind of animal she was—as opposed to the kind of animal I thought she ought to be. When I consider our exchange, I think we about broke even. She liked living in my house because I offered her food, a safe territory, and a playmate for her kittenish moments. And I liked having her indoors because she enriched my home immeasurably with her distinctive animal presence.

FLEAS

*H*ussy introduced me to another life form that thrives indoors: her fleas. She had been a stray before I met her and had probably played host to these parasitic, or dependent, insects for as long as she could remember. But after they worked me over the first night she spent indoors with me, I put a flea collar on her and rid her body and our household of her pests.

She didn't have fleas again, or at least she didn't share any of them with me, until we moved to Vermont. This second outbreak taught me how clever fleas can be. We had been in the farmhouse that was to be our new home for only a few days when I noticed Hussy scratching and biting at herself. I was so busy settling in that I ignored her behavior—until I woke up with flea bites once again.

I couldn't imagine where she had gotten this new batch of fleas. She hadn't had them when we left our old home, and she had been nowhere except in the car or in our new home since then. The fleas, it turned out, came with the farmhouse. They had been waiting patiently for a new cat or dog to move in because their former host, a large dog belonging to the family who sold the house to me, had moved out on them.

Fleas are accustomed to shifting from one host to another. They are somewhat inconvenienced by the total disappearance of available hosts—as during the month my new house was unoccupied—but they are not necessarily destroyed by such circumstances. Adult fleas can last for several weeks or months without a meal of blood, and because the earlier life stages don't need blood at all, a whole generation can grow up waiting for a new host to move in.

A flea's eggs drop from a cat's or dog's fur and land in the animal's bedding or on the floor. When the eggs hatch a few days or a few weeks later (depending on the temperature), the young, called larvae, feed on particles of organic matter in dust and debris. When the larvae are fully grown, they withdraw into a cocoon to pupate. The duration of the pupal stage, an inactive, nonfeeding period during which the larva *metamorphoses* into an adult, is variable, allowing for a response to conditions such as those encountered by my farmhouse fleas.

If a larva enters pupation without a warm-blooded host around to inhabit as an adult, it matures but stays inactive inside its cocoon until it feels the vibrations—the footsteps, perhaps—of a potential host. Only then will it emerge from its inactivity, an adult flea equipped with bloodsucking mouthparts and ready to produce another generation of fleas. My cat inherited the dog's fleas who had, one way or another, managed to survive the month without a host.

Most species of fleas have their favorite hosts—dogs, cats, rodents, or human beings, for instance—but they are also practical enough to settle for what's available in the way of warm blood. Fleas can't live on just any warm-blooded animal, however. They must pick host species that have a way of life compatible with their own. An adult flea likes to feed, drop to the ground for a while, then hop back onto its host—or onto another one—when it's ready to feed again. Because fleas also spend the early stages of their life cycle away from the host's body, they need a host that will still be around when they mature several weeks or months later. The animals that fleas parasitize, therefore, are animals that nest, live in burrows, or sleep in regular spots. Wandering animals like deer don't have fleas. Nor do apes and monkeys. Human beings, with their settled ways, are, in fact, the only primates to have attracted their own species of fleas.

Fleas are well adapted to the parasitic lives they lead. They began to evolve toward their present life-style long ago, when some of the animals around them began to evolve toward warmbloodedness. The insects that became fleas responded to the new circumstances by establishing a place for themselves among newly evolved feathers and fur. But the parasitic habit brought some changes. Fleas lost their wings, for instance, because they no longer needed to fly. They still needed to get from ground to animal or from one animal to another, however, so they developed an alternative method of locomotion—one for which they are justly famous. Fleas can leap 7 or 8 inches (17.5–20 cm) into the air or 12 inches (30 cm) across the floor. A human being would have to high-jump 330 feet (100 m) or broad-jump 990 feet (300 m) to match this performance.

The shape of a flea's body has become tailored to its habitat. It is slender—almost flat—vertically, which enables it to crawl between its host's hairs. The spines on its body slant backward, both to ease its movements forward through the hair, and to catch in the hair when the host animal scratches or bites at it. The flea's outer covering is tough and smooth and therefore difficult for an animal to hold onto with claws or teeth.

It's easy to think of fleas as nothing but small and irritating pests to be gotten rid of as soon as they appear. But if your household pet presents you with them, you might seize the opportunity—as I did the second time around—to examine some of their adaptations with a hand lens before you follow your veterinarian's advice on how best to eliminate them. Fleas are actually impressive little survivors, who manage to live quite well off the warm-blooded animals that evolved around them.

DOGS

While I've shared my various homes with only one cat, I've lived with four very different dogs. The first was an English setter, the second a boxer, the third a stray beagle, and the last a Humane Society mix—German shepherd and collie as far as I could tell. As my own experience indicates, domestic dogs come in so many different sizes, shapes, abilities, and temperaments, it's impossible to describe a typical one. But if you observe one individual dog closely—your own pet or a dog that belongs to a neighbor or friend—you can perhaps figure out the genetic route that particular dog has traveled as human beings bred its type farther and farther away from the wild wolves who were its ancestors.

The exact process by which the first wolves became associated with human beings is not known, but it happened at least fourteen thousand years ago. Then, sometime after these wolves became human companions, wolf owners began to select individuals who were especially good at one thing or another and breed them with other individuals who were good at the same thing. This selective breeding gradually produced all the different breeds of domestic dogs we see around us today. To understand what kind of genes—the elements that control hereditary traits—have been encouraged in the dog you happen to be focusing on, you need to know what its ancestors were originally bred to do. Conveniently, the American Kennel Club divides American breeds into seven different classes based on the various roles dogs are—or were—intended to play.

Dogs bred to help human beings with hunting account for three of these classes, some with certain subgroups. One class includes all the hound breeds, with subgroups for sight hounds and scent hounds. The ancestors of sight hounds, such as the long-legged, fast-running greyhounds, Salukis, and Afghans, were selected for keen eyesight and speed. These breeds are called sight hounds because they follow their prey with their eyes. The ancestors of the scent hounds, who produced short-legged sniffers, such as beagles, bassets, dachshunds, and bloodhounds, were selected for their sensitive noses. Short legs were desirable in these breeds because their noses would be closer to the ground. Scent hounds can smell even the faintest odors of skin secretions left behind in animal tracks.

A second class of hunters includes all the terrier breeds, whose ancestors were feisty, aggressive, persistent, and without fear of the animals they chased, cornered, and harrassed. They were bred to rout animals such as foxes, badgers, ferrets, and rats out of burrows in the earth, or *terre*. Familiar terriers include Scotties, Airedales, and fox terriers.

Sporting breeds make up the third and last class of hunting dogs. Their ancestors showed a preference for serving and obeying a human master over dashing about on doggy missions. This desire to serve made them useful adjuncts in the delicate and disciplined business of hunting birds. The sporting class includes three subgroups: the pointers and setters, who stand still and point their masters to a hiding bird; the spaniels, who flush hiding birds for their masters' convenience; and the retrievers, who plunge into water to carry shot birds back to their masters, without indulging themselves in even a little bite.

Human beings also needed help with specific work tasks, which led to the selective breeding of working and herding dogs. Some of the workers were bred for strength, producing such powerful pullers as huskies and malamutes, while others, such as the ancestors of mastiffs and Doberman pinschers, were bred for their intimidating protectiveness of territory and family, which makes them effective guard dogs. Herders were bred for their ability to move and protect sheep, producing such solicitous shepherds as collies and German shepherds.

The sixth and seventh American Kennel Club classifications reflect more about human tastes—and the domestic dog's genetic plasticity—than human needs for assistance. The ancestors of the toy breeds were selected for their diminutive size and have produced both miniature versions of larger breeds and little lap sitters such as Chihuahuas and Pekinese. Finally, the dogs classified as non-sporting breeds have been developed primarily for companionship or competition in dog shows. The best known of these is the pampered poodle.

Not all breeding, of course, has been selective. Many dogs have interbred with dogs they chose for themselves rather than the mates their owners might have intended. And some owners haven't cared as much about selectively bred characteristics as general friendliness, which mixed breeds sometimes inherit quite accidentally. Even these mongrels betray evidence of their ancestry, however. You just have to be more of a detective to perceive it.

Despite the many differences in size, shape, ability, and temperament, all domestic dogs share certain behaviors with their common ancestor, the wolf. The most important behavior—the one that invited some wolves into human groups to start with—is sociability. Like a wolf, a domestic dog needs a social life in order to develop into the kind of animal it's intended to be. If a pet dog is going to be a happy and well-adjusted member of a human family—which becomes it surrogate wolf pack—it needs companionship, approval, acceptance, discipline, and a clear sense of its status within the family pack. And most of all, it needs strong leadership from its owner, who functions psychologically as this domestic dog's head wolf.

GUINEA PIGS

You don't have to own every animal the human species has ever welcomed indoors to observe it. Pet stores offer concentrated assortments of indoor animals free for the looking—in hopes, of course, that you will buy one. If you already know that you don't want another animal in your home, strengthen yourself before you enter a pet store because the creatures you will see are living proof of their species' ability to meet human needs—or to create new ones.

The guinea pig, a stout, furry little animal that looks somewhat like a rabbit with shorter, floppier ears, smaller eyes, shorter back legs, and no tail, is the first rodent to have won a place in human homes. Rodents are gnawing mammals with strong, perpetually growing front teeth that must be kept sharpened and shortened by chewing on hard things. In the wild, rodents gnaw on hard-shelled seeds (like a squirrel on acorns), or on woody plants (like a beaver on trees). In a human home, food containers and furniture become candidates for rodential gnawing, which explains in part why pet rodents must be kept in cages. Also, rodents cannot be litter-trained or housebroken, so keeping them caged is the only way to localize their excretions.

Guinea pigs are an exception among the domesticated rodents, which include hamsters, gerbils, and special breeds of rats and mice, in that the early stages of their domestication did not involve cages. Nor did it involve their potential as laboratory animals or even their appeal as pets. Guinea pigs first entered human dwellings—the shelters of pre-Incan inhabitants of Peru—as food.

Archeologists speculate that wild guinea pigs may have contributed to their own domestication by seeking out burrowlike living quarters within the warm, dry spaces human beings created for themselves. In the wild, guinea pigs don't dig their own burrows but rather crawl into natural rock crevices or burrows abandoned by other animals. The pre-Incan residents of Peru would have welcomed wild guinea pigs into their homes—and perhaps even created burrowlike structures to invite them in—because these plump, prolific little rodents constituted a major portion of the local food supply. Having them living and reproducing indoors meant a readily available and self-perpetuating supply of meat, which before had required chasing and trapping in grassy fields. Even today, guinea pigs in remote Peruvian villages have the free run of human kitchens, living and breeding under loose supervision while they're waiting to be eaten.

The exact date of the guinea pig's first entrance into human dwellings is uncertain, but archaeological evidence indicates that they were domesticated by 3000 B.C. By the time Pizarro conquered the Incas in A.D. 1532, domesticated guinea pigs, which the Incas called *cuys*, were being traded, bred, and eaten throughout much of South America, and shortly thereafter they were transported—momentously—to Europe.

The Europeans perceived these furry little rodents, which for reasons that mystify language historians, they named guinea pigs, as perfect pets. Just as the guinea pigs' burrowing habits led them into South American homes, where they were eaten, their social habits made them welcome in European homes, where they were caressed, cuddled, groomed, and even displayed in competitive shows—for which occasions they were called *cavies*, a shortening of their Latin name, *Cavia porcellus*.

Guinea pigs adapted readily to their new identity because

they are gregarious animals. Wild guinea pigs live in colonies and seek physical contact with other members of their group for comfort and security. In a human home, a guinea pig adopts the family as its social group and responds happily to lap sitting, hugging, and petting. In fact, if a solitary guinea pig doesn't get enough physical attention, it becomes lonely. Consequently, many households wind up with pairs or groups of guinea pigs.

But groups of guinea pigs can be tricky because of their mating habits. The guinea pig's natural group consists of a single male living with a harem of several females, with whom he mates as often as they will let him. Unregulated behavior of this type—even if the male's harem consists at the outset of only one female—can soon lead to a colony of guinea pigs too big for a human home. Therefore, a male and a female aren't a good combination, and two males will fight, which leaves only two or more females for a socially compatible group.

Late in the 1700s, about two hundred years after these edible rodents had chanced into their new career as pets, they became subjects for laboratory experiments. While it was their appearance and behavior that made them appealing pets, it was their biology that made them appealing laboratory subjects. Guinea pigs are strangely like human beings in some aspects of their biology, including their need for vitamin C from a source outside their own bodies, their production of well-developed, open-eyed young, and their susceptibility to human diseases. Guinea pigs have been used in so many experiments involving vitamin C, heredity, genetics, diseases, serums, and drugs, that their name has become synonymous with "experimental subject"—as in "human guinea pig."

Guinea pigs have had a long and evolving association with the human species, which has included different names and different identities in different times and places. While their own habits and behaviors may have brought these rodents indoors in the first place, human habits and behaviors have shaped the variety of experiences they've had there.

HAMSTERS

The hamster, also a rodent, has a much shorter history of association with human beings than does the guinea pig. The individual destined to become the ancestor of all of today's pet hamsters was still living in the wild in 1930. That year, a zoologist from the Hebrew University in Jerusalem was collecting native animals in Syria and happened upon a burrow belonging to a golden hamster. European and Chinese hamsters were already well known, but these golden hamsters, which occupy a limited range around Aleppo, had not yet been studied. The professor dug eight feet into the hamster's extensive burrow before he finally found the occupants—a female hiding in her nest with a litter of young. He transported this family group back to Jerusalem, where some escaped, some died, and three—a male and two females—survived to produce the first golden hamsters born in captivity. Within a year, offspring of this original family had found their way into medical laboratories.

Golden hamsters, who have soft fur, short tails, and look somewhat like small bears, were an instant hit as laboratory rodents. They were smaller than guinea pigs, less odiferous than rats and mice, and faster breeding than both. They were first used in studies of a disease that strikes Middle Eastern children and have since been involved in studies of tuberculosis, leprosy, cancer, and even tooth decay.

But hamsters had one shortcoming as laboratory subjects—they bit when alarmed. Researchers soon learned, however, that if they handled young hamsters gently and played with them, they became responsive, trusting, and could even learn tricks. Shortly after World War II, these attractive little rodents moved from the laboratory into human homes and became exceedingly popular pets.

Like a pet guinea pig, a pet hamster must live in a cage to protect its human home from gnawings and excretions, but its habits are otherwise unlike a guinea pig's. Whereas a guinea pig is social, thriving in a cage with others of its own kind, a hamster is solitary. An individual hamster can learn to enjoy being handled by human beings, but if you try to add another hamster to its cage there might be trouble.

In the wild, female hamsters defend small territories against both males and other females, while males wander in hopes of chancing through a female's territory just when she's ready to mate. A female will tolerate a male—and perhaps even offer him some food—on the day she's ready to mate, but if he arrives too soon or stays too long, she will attack and maybe even kill him.

Because the female is so aggressively territorial and the timing of her receptiveness so exact, consisting of only an hour or so every fourth day, professional hamster breeders have learned to put the female into the male's cage when they think she's receptive rather than the other way around. Outside her own territory, the female is less antagonistic, but still, the breeder has to watch the pair's interaction closely. If they don't warm to each other, the match might have to be delayed until the next evening—the female's heat usually occuring at night—or until the next time the female reaches the exact point in her four-day cycle when she's ready to mate again.

The female hamster is also aggressive toward young hamsters and is very likely to eat them except during the three weeks her hormones are telling her to care for her own offspring. But she is a very nervous and sensitive mother and might eat even them if something upsets her. Therefore, hamsters are best kept alone in their cages as solitary pets, with breeding left to professionals who understand the subtleties of hamster breeding.

Guinea pigs and hamsters differ also in their daily patterns of activity. In the wild, guinea pigs are *crepuscular* animals, with most of their activity occurring at dawn and dusk, but they respond to indoor living conditions by becoming *diurnal,* or active throughout the day. Hamsters, in contrast, are nocturnal, tending to hide and sleep all day and to become active at night.

Other behavioral differences between guinea pigs and hamsters reflect the different environments where they evolved. The ancestral guinea pigs found plenty of food for themselves throughout the year, so pet guinea pigs take for granted the food you offer them, nibbling at it whenever they feel inclined. Wild golden hamsters, in contrast, compete for a limited, seasonally available food supply. Like some North American mammals, they hibernate to survive cold weather. A pet hamster won't hibernate at normal indoor temperatures, but it will hoard food to keep it from an imagined competitor and to have it on hand in case the weather changes. Its name, in fact, comes from the German word *hamstern,* which means "to hoard" or "to store." You can watch a hamster stuff its expandable cheek pouches full of seeds and carry them to a hiding place somewhere in its cage. When you clean the cage, you will notice that this orderly little animal has designated different parts of its cage, which is the indoor equivalent of its burrow, for its stored food, its toilet activities, and its sleeping quarters.

Unlike most other animals that have become pets, none of the wild hamster's needs or habits inclined it toward an association with human beings—except as an agricultural pest. But the descendants of the fated female who was dug out of her burrow near Aleppo seem to be making the best of their unsolicited condition.

GERBILS

*H*amsters were the rage during the 1950s, but their nocturnal habits and their tendency to bite when alarmed made them vulnerable to competition from a similar-sized rodent that arrived on the pet scene a decade later: the Mongolian gerbil.

This long-tailed, big-eyed little rodent, which looks like a small kangaroo, found its way into human homes much the same way the hamster did. It all began in the laboratory. In the 1880s, a Russian scientist trapped gerbils in the Mongolian desert to study tuberculosis. Other researchers in Europe and Asia experimented with gerbils throughout the first half of this century. But it wasn't until 1954, when twenty-two Mongolian gerbils were sent from a Japanese supply house to a laboratory in the United States, that these Asian rodents began their rise to popularity as American pets.

Like hamsters—and unlike laboratory rats and mice—gerbils

are clean, odorless, and conservative in their excretory habits. Other characteristics that endeared gerbils to laboratory workers include their alertness, curiosity, gentle dispositions, and responsiveness to human touch. And finally, gerbils are active during the day, as are most of their human observers and attendants.

Gerbils and hamsters are members of the same family, but they reflect entirely different environmental adaptations and social patterns. They both dig burrows, but the species of gerbil that has become the pet store favorite excavates deserts in Mongolia rather than farm fields in Syria. While hamsters have special cheek pouches to help them store seasonally available food, desert-adapted gerbils have internal adaptations to help them conserve water. In the wild, gerbils get the water they need from the stems and leaves of the plants they eat, and they urinate sparingly to keep the water in their systems. Their droppings, too, are very dry, making gerbils quite easy to clean up after.

Gerbils are also much more sociable than hamsters. Adult gerbils live in monogamous pairs within a larger colony. The males establish territories within the colony, but once a female has chosen a male, she defends the territory, too. Both sexes participate fully in raising the young, and the young are permitted to stay with the family group as long as they like. They are not, however, permitted to breed until they disperse and establish territories of their own.

Unlike the other small rodents we keep as pets, gerbils are happiest as mated pairs. But a pair of gerbils, like any two rodents of opposite sexes, will overpopulate a cage in short order. The social dynamics that operate within a family of gerbils would keep the young from mating, but the parents themselves can produce five litters, or twenty to twenty-five offspring, in a year, which would crowd even the largest cage and stress even these most amicable of rodents. So perhaps gerbils, like guinea pigs, are best kept as groups of females, who won't fight over territory as long as males are not around, or as single animals, with lots of attention from their human keepers to provide the social interaction they need.

As you observe the caged rodents side by side in a pet store, you will notice differences in appearance, behavior, social interaction, and attitude toward you, which will teach you something about these animals. And if you're at all like me, you'll also learn something about yourself. During the hour or so I spent rodent-watching one afternoon, the gerbils were tearing around their cage, which was a large aquarium, pawing frantically at the glass that separated them from me, and then digging up their bedding as if they were trying to tunnel their way out. I must admit, I found their apparent interest in me quite flattering.

Meanwhile, most of the hamsters were fast asleep. But one woke up and started grooming itself, licking its tiny pink paws and pulling them over both ears simultaneously. The small creature's fastidiousness and soft, dense fur tempted me to reach into the cage and cradle it in my hand despite its self-absorption and total indifference to me.

But the guinea pigs were the most engaging of all. They are bigger than mice and have blunter noses than rats, so that I didn't have to overcome any initial repugnance, and their fat, moplike bodies invited touching. They mostly jostled, snuggled, and squealed as they looked out at me, but when one suddenly performed what looked like a little jump for joy, I had to laugh out loud.

I'm not sure that watching caged rodents in a pet store is the best training for an aspiring naturalist, but one of my theories is that the more I learn about the animals that have won their way indoors, the more I will know about the human species. And understanding the human species is one of my long-term goals—in the hope that I may someday learn how we ourselves might fit back in the outdoors.

CANARIES

Canaries look like technicolor versions of some of the small seed eaters who hang around bird feeders during the winter. If you can ignore their brilliant colors, which these days include more shades than just canary yellow, other features will indicate their basic kinship to your feeder birds. The canary's feet mark it as a passerine, or perching bird, a large group that includes most of the familiar birds that visit feeders during the winter or sing songs and nest around yards during the spring and summer. Perching feet are designed to grasp branches, three slender toes pointing forward and one wrapping around from behind. The canary's strong, conical little beak further identifies it as a member of the finch family, making it a close relative of goldfinches and sparrows.

Canaries have been popular for so long it seems they've always been with us, but they became indoor cage birds only slightly before guinea pigs arrived on the European scene—both having been discovered in different parts of the world by Spanish explorers. The ancestral canaries inhabited the Canary Islands, just off the northwest coast of Africa. These islands were perfectly situated as a stopover for European ships embarking upon voyages of discovery, and the Spanish and Portuguese fought over them during the 1400s. In 1479 they fell permanently to Spain, giving the Spanish a monopoly on the canary trade that developed later.

Exactly how the canary trade got started is not known, but at some point the songs of the island's wild finches entranced Spanish sailors, who took cages of them back to Europe to sell. One historian theorizes that the pre-Spanish inhabitants of the Canary Islands, known as Gaunches, had already been keeping wild canaries in primitive cages, which might have given the Spanish sailors their enterprising idea.

The wild canary's song was what first attracted human interest, but after the species had been bred in captivity for several generations, some individuals began to show unusual colors, feather textures, and body shapes. The wild bird is streaked olive green with some yellow on its underparts, sides, rump, and head. Individuals who showed more and more yellow led to the development and subsequent popularity of the all-yellow bird most of us picture when we think of a canary.

Although breeders have continued to experiment with color, feather texture, and body shape, song remains the essential feature in canaries that are kept as household pets. It's difficult to describe a typical canary's song because selective breeding has led to two types of singers. The types called choppers sing loud, clear songs that closely resemble the songs of the wild birds. Their name comes from the "chop-chop-chop" sounds that can be heard among their notes. Choppers are popular as pets because they sing spontaneously with their beaks wide open. The types called rollers sing a softer, more continuous, and rolling song. They hardly open their beaks at all as they perform certain notes called rolls and tours. Rollers must be taught to sing in their special way and are often trained by older birds, who provide polished songs to imitate.

Even though canaries have been selectively bred for more than four hundred years, they are still subject to the biological laws and seasonal variations that governed their wild ancestors. For instance, it's still only the males who sing, and they still sing most intensely during their nesting season. An enterprising researcher discovered that female canaries can be induced to sing, too, if they are treated with the male sex hormone. But when the hormone wears off, the singing stops. Because males bring higher prices than females, and the sexes are almost impossible to tell apart from external appearances, dealers have been known to treat their females and pass them off as singers. But most breeders and pet store owners today do their best to determine that the young birds they sell as singers are indeed males, even offering to replace the bird if it doesn't sing.

Other research has revealed that male canaries learn their songs anew each year and can change their songs from year to year in response to new songs they hear. The parts of the male's brain responsible for making him sing actually increase in size as he gets ready to mate and then shrink again when the nesting season is over.

The singing of caged canaries is at its best from fall, when males begin practicing, until early spring, when they are singing in earnest in hopes of impressing a female. By midsummer, the males have become quieter, and they remain completely silent during the molt, or shedding of feathers, which lasts from late summer until fall.

In the wild, this annual shedding of worn feathers and regrowing of new ones keeps canaries in good flying condition. Different species of birds molt in different ways, but canaries molt as most other perching birds do. They shed their feathers gradually so that they never lose their ability to fly, unlike ducks and geese, who shed their flight feathers all at once and become flightless for part of the summer. During the six to eight weeks canaries are molting, they can look fairly bedraggled, but by fall they have finished growing in their new feathers and are ready to begin singing their way toward another nesting season.

Although we think of canaries primarily as music makers, they are not like music boxes, radios, or record players. We can't just turn them on and off, expecting them to sing continuously or whenever we want them to. Because they are living creatures, they sing in accordance with their own seasonal cycles and biological drives, merely including us in their audience because we've located them in our homes.

PARAKEETS

Parakeets are about the same size as canaries, but if you look closely at the two species, you will see that they are very different birds. The parakeet has a strongly hooked beak and manipulative, climbing feet with two toes pointed forward and two back. Instead of sitting on its perch, as a canary does, a parakeet is more likely to be walking right up the side of its cage, using its beak and both feet to provide three points of leverage and support.

Parakeets, as their beaks and feet indicate, are closely related to parrots, but they entered human households by a completely different route. Whereas the best known of the larger talking parrots came from the jungles of Africa and South America courtesy of sailors, parakeets came from the grasslands of Australia courtesy of an English ornithologist.

The first pet parakeets, which the English call budgerigars

or budgies, from the native Australian word *betcherrygah*, meaning "pretty bird," arrived in England in 1840. John Gould, the ornithologist who brought them, had been observing, collecting, and painting Australian birds and decided to bring some living specimens of these canary-size parrots home with him. They were instantly popular and became increasingly so when Europeans discovered that they could, like canaries, be selectively bred to produce a variety of different colors, and could also, like the bigger parrots, be trained to talk.

Because pet store parakeets have been selectively bred for so long, very few display the species' natural coloration. The wild Australian bird is basically green to blend in with its grassland habitat, but it is also distinguished by a bright yellow face, fine black barring down its back, blue cheek patches, black spots on its chest, and a long, dark-blue tail.

Other markings differentiate young from mature birds and males from females. A young bird, which is what you want to buy if you have fantasies of teaching your pet to talk, has black barring on its forehead. This barring disappears when the bird molts at the age of three months. Another feature that will change—and might already be showing signs of change in young birds—is the cere. The cere is the patch of thick, bare skin where the bird's beak joins its forehead. In adult males it turns blue, while in adult females it turns brown. In young parakeets of both sexes it is pink, sometimes with a tendency toward either the blue or brown it will one day become. Some experts claim that these sexual differences matter because the males make better talkers, but other experts brag about their talking females. Perhaps canaries have created an unnecessary prejudice against female parakeets because among canaries, of course, only the males sing.

In the wild, parakeets travel in huge flocks, wandering nomadically from one area to another as wild grasses produce their seeds. Because they live in a dry environment, their breeding cycle is influenced less by seasons than by rain. Unlike canaries, which can breed only during the spring, parakeets are capable of breeding throughout the year. Normally, they seize the opportunity of a heavy rain, which assures them a good supply of freshly ripening grass seeds at about the same time their eggs will hatch.

Like other members of the parrot family, parakeets mate for life, so the males do not have to sing brilliant songs each year to win their females. Their potential for speech, however, is related to the way both sexes work to maintain their lifelong pair-bond. During quiet moments, when the pair is separated from the larger, chattering flock, they sometimes mimic each other's sounds to reinforce their special one-on-one relationship.

The parakeet that becomes a pet merely transfers its bonding behavior from a member of its own species to a human companion. It learns to imitate the words and phrases it hears repeated over and over because it wants to please its trainer, whom it has come to perceive as its life partner. If, in return, the imitated words inspire the trainer to offer food, attention, or affection, the parakeet has ample reason to develop a considerable vocabulary.

For this cross-species relationship to work, the parakeet must bond with a human being while it's still young. At the point in the nesting cycle when wild young parakeets would have left their parents, captive-bred young parakeets are separated from their parents and offered for sale. These six-week-old birds are just ready to form their social attachments, and if one is isolated from other members of its own species and taken into a human household, it will adopt the family as its flock and the person who spends the most time with it as its mate.

But this transition doesn't happen automatically. The young parakeet must learn to feel safe in its human home, and it must learn to trust its trainer. The bird will talk only if it feels completely comfortable, and even then, the process might take four to seven months. A parakeet's voice is smaller and higher-pitched than a parrot's, but its repertoire of words, phrases, and rhymes can be as extensive if the individual bird is properly motivated.

While our other popular pets have appearances, habits, or behaviors that make them appealing to watch, touch, and listen to, parakeets go one step farther. If we meet their needs, they will reward us with fragments of our own language.

AMERICAN CHAMELEONS

Reptiles aren't as popular as birds and mammals, but they, too, have their devotees. Of the reptiles that have found their way into human homes, one of the most common is the 5- to 7-inch (12.5- to 17.5-cm) lizard known as the American chameleon—named after but not closely related to the more difficult to care for Old World lizard known as the true chameleon. Even I, who as an adult have developed an embarrassing reptile phobia, once owned an American chameleon.

I bought it at a circus when I was eight years old, in large part because it looked like a miniature version of the huge prehistoric reptiles that fascinated me on school field trips to the natural history museum. The salesman merely clinched the sale by promising me the chameleon would change colors. I obviously had no reptile phobia then, but my new pet escaped from the shoebox I tried to keep it in before I had a chance to learn enough about it to mitigate my future fear.

My pet escaped, sparing itself the unintentional abuses of an eight-year-old keeper, because I had no idea how to contain, let alone care for, a chameleon. An American chameleon is best kept in a terrarium—an aquarium that's been set up to resemble a terrestrial, or dry-land, habitat—with a screen across the top. The screen is necessary because a chameleon can walk right up a pane of glass. Being a tree dweller, it has sharp claws for clinging to rough bark, but it also has sticky toe pads that adhere to smooth twigs and leaves. My cardboard shoebox wasn't even a challenge for such an accomplished climber.

Chameleons climb in search of insects. If I had managed to detain my pet long enough for it to need food, I probably would have starved it to death on the diet I planned to offer it. I would have tried to make it drink milk from an eyedropper or eat hamburger off the tip of a toothpick. Or maybe I would have offered it some breadcrumbs, a bit of my dessert, or other human tidbits. But a chameleon has no idea what to do with such inanimate foodstuffs. It is adapted to hunting live insects, following slight movements with its separately focusing eyes until it's close enough to use both eyes together for depth perception. A chameleon stalks an insect much as a cat stalks a mouse, freezing when it's close enough to strike and then pouncing quite suddenly to catch the unsuspecting fly, beetle, moth, spider, cricket, grasshopper, or caterpillar in its wide-open mouth.

If my poor captive hadn't starved to death waiting for me to provide it with live insects, it certainly would have died of thirst. A chameleon doesn't know how to drink water out of a dish, even if the water is warmed, sweetened with sugar, or otherwise made tempting to a water-drinking animal. In its native habitat—the moist woodlands of the southeastern United States, where this familiar little lizard is known as the green anole (uh-NOH-lee)—it gets the water it needs by absorbing moisture through its skin or licking rain and dewdrops off leaves. So a well-cared-for pet chameleon needs an environment that's considerably moister than a shoe box, an environment that can be created by including some live vegetation in the terrarium and misting the vegetation daily.

Provided with a regular supply of food and moisture, a captive chameleon can live comfortably, if somewhat uneventfully, indoors. In the wild, a chameleon has to contend with predators, competitive members of its own species, and prospective mates. These activities involve stress, which is what a chameleon's famous color changes are all about. The chameleon's most dramatic color changes are not attempts at camouflage, but reflections of its emotional state. When the animal is basking in the warm sun, completely relaxed, it is usually brown, but if it is alarmed, it will turn bright green.

If the cause of alarm is a predator that has grabbed it by its long tail, the chameleon can shed the tip and flee. The tail tip will twitch independently for up to five minutes, leaving the hungry predator busy trying to subdue this expendable part. Meanwhile, the frightened chameleon can find a place to hide, and over the next few months it will grow a somewhat shorter tail tip to replace the one it lost.

In encounters with other members of its own species, a chameleon must be alert to the appearances and behaviors that identify sex and status. Adult males are territorial during the breeding season and will not tolerate the presence of competitive males, but they will tolerate submissive males and as many females as care to feed—and mate—within the territory. A male advertises himself and his territory by nodding his head, bobbing up and down, and showing off a bright pink flap of skin—called a dewlap—beneath his chin. A competitive male displays back, which might lead to a fight, while a female either dashes away or responds to his performance by standing still and allowing him to mate with her. Females, meanwhile, compete with other females over good feeding spots and are, in fact, more interested in feeding than mating most of the time. A female produces her eggs singly—one about every two weeks—and is unreceptive to males between eggs.

Watching a captive chameleon lacks the challenge of observing one in the wild. But then again, examining the animal itself, which is much easier to do when it's leading a relaxed and protected life, offers an opportunity to understand how its physical features relate to its life in the wild. If the chameleon I bought at the circus had given me enough time to figure out how to provide for it, perhaps I would have developed an abiding affection for reptiles instead of my embarrassing phobia.

GOLDFISH

Keeping fish in indoor aquariums is a relatively recent development, but the goldfish bowl dates back to the 1330s, and the goldfish itself to about the year 1000, when Chinese aristocrats began collecting naturally occurring "golden fish" and breeding them in outdoor ponds. These colorful fish belong to a species of carp native to the shallow, weedy ponds and slow-moving streams of eastern Asia. In the wild, this species is usually a shade of grayish olive that blends in with its natural surroundings, but this muddy color is actually a mixture of black, orange, and yellow pigments. Occasionally, a wild individual is born without the black pigment, resulting in an orange-yellow fish.

At first these attractive fish were kept only by the wealthy, who maintained special goldfish ponds, where they selected and bred the individuals with the best colors. But by 1330, goldfish had become popular in the city of Peking, where those who could afford to kept them indoors in jade vases, and more average folk kept them in simple bowls. In the six hundred fifty years since goldfish hit Peking, they have traveled to Japan, where they were bred not only for color but for unusual shapes, to Europe, where during the 1800s fashionable women wore them dangling as earrings in miniature glass bowls, and to North America, where shortly after the Civil War they became popular pets.

A goldfish in a bowl offers a convenient introduction to fish—the way they move, breathe, and live in perfect accord with their aquatic medium. If you watch a goldfish swim, for instance, you will notice that fish have the attributes of both snakes and jets. The goldfish wiggles its lithe body as if it were an extremely active snake, propelling its streamlined shape forward as if it were an extremely maneuverable jet.

The large tail fin provides the main thrust, while the other fins serve to turn, lift, stabilize, and stop the fish. The pair of fins near the head—those that are the counterparts of human arms—provide direction. The pair on its belly elevate the fish, but they also roll it slightly from side to side. This tendency to roll is corrected by the fin on the fish's back and the one just in front of its tail. If the fish wants to stop, it uses its "arm-fins" as brakes, and if it wants to rest in a stationary position, it flaps these same fins rhythmically to counteract the rhythmic spurts of water coming from the openings on the sides of its head.

As you watch a goldfish hovering gracefully in one spot, you will see that it seems to be mouthing the words, "yup, yup, yup." This repeated opening and closing of the mouth is the way a fish breathes. It draws in some water, which passes over the gills—delicate, thin-skinned fringes filled with small blood vessels. Oxygen from the water can move right through the thin skin directly into the blood, which transports it to the rest of the body. Wastes, such as carbon dioxide, can move the other way, from the blood through the thin skin into the water, then out through the gill slits, the openings that show quite visibly on the sides of the goldfish's head.

Everything else about the goldfish is adapted to its watery existence, too. Overlapping scales lubricated with mucous enable it to move efficiently through water. Its large, lidless eyes, which researchers speculate are nearsighted, enable it to see far enough to detect food or danger, and its nose, which is totally unrelated to its breathing apparatus, enables it to extract scents from the water that washes through its small nostrils.

Goldfish have become popular indoor pets because they adapt more readily to the vagaries of human care than do most other fish. Although a goldfish appreciates a filtered and aerated aquarium, it can survive in a bowl of tap water as long as the water is fresh, and not too soft, hard, or chlorinated. It tolerates room-temperature water—even if it prefers water that's cooler than most rooms—and will eat almost anything you feed it. A goldfish is an omnivore—an eater of a wide variety of plant and animal foods—who needs regular feedings of commercial goldfish food to balance its diet but will also eat bread crumbs and cereal (which are not especially good for it), and small pieces of earthworm, insects, and bits of plants.

Because two or more goldfish kept together in a bowl will stay relatively small in response to their limited space, you don't need to worry about their gender. Goldfish don't mate until they have attained a size of about 3 inches (7.5 cm), and even then, a pair would probably not be inspired to mate within the confines of indoor quarters. Commercial breeding involves fish living in carefully managed outdoor ponds, and many of them are more than 10 inches (25 cm) long.

Goldfish young do not hatch from their eggs looking exactly like their selectively bred parents. At first they sport the grayish olive color of their wild ancestors, developing their parents' traits only several months later. Even then, only a few of the offspring will display the brilliant colors or unusual shapes of their parents. Goldfish breeders figure that for each one hundred eggs that are laid, they might have six to eight goldfish to sell.

The genetic variability of goldfish is demonstrated most impressively, however, not when they are bred commercially, but when some individuals escape—or are released—into the wild. Over just a few generations, their descendants will revert to the natural grayish-olive color and functional shape of their ancestors, fitting right back into their new environments as if they had never taken a detour through the ponds and fishbowls of their human admirers.

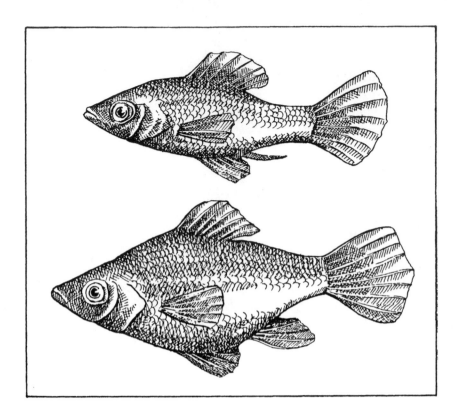

GUPPIES

Goldfish are so hardy and adaptable that they had spread to fish-bowls all over the Orient, Europe, and North America by the time guppies arrived on the scene. Guppies, which come from the warm rivers of the Caribbean islands and northern South America, had to await the development of heated aquariums before they could move into human homes.

The timing was perfect. During the mid-1800s, European fish enthusiasts developed effective methods for heating water in glass containers. Public aquariums full of exotic tropical fish began opening in major cities, and the small, home-size aquarium wasn't far behind. Amid this wave of public interest in tropical fish, the Reverend Robert J. L. Guppy, an Anglican minister in Trinidad, discovered some small fish with bright markings swimming in a local river and sent some home to the British Museum. These first guppies, named for the good Reverend, arrived in England in 1866, and by 1908 they were being promoted commercially as the perfect beginners' aquarium fish.

Guppies make good aquarium fish because, like goldfish, they are adaptable and can tolerate the wide range of conditions they encounter under human care. Furthermore, they offer home aquarists something that goldfish don't: They breed prolifically indoors. Their offspring are so variable that new lines can be selectively bred and entered in the international shows that fuel the increasingly popular tropical fish hobby.

If you observe the guppies that are available in pet stores today, you will see the direction selective breeding has taken. The guppies I saw had brilliant red tails, some streaked with blue. Some of the guppies had other colors along their sides, and as I continued to study the individual fish I discovered that each one seemed slightly different. When I stood back to look at the aquarium as a whole, however, I saw a tank full of bright, rapidly moving colors.

The potential for these colors resides in wild guppy genes, but selective breeders have developed them beyond the colors evolved by wild fish in their native rivers. In the wild, only male guppies are colorful, the females tending toward the grayish olive that serves as effective camouflage in aquatic environments. Guppy males, which are smaller than the females, must display themselves boldly to attract a female's attention, and the more unusual a male's coloration, the more attractive he is to a prospective mate. Therefore, wild males are both colorful and variable, which has provided selective breeders with ample raw material. The only problem with some of our modern pet store guppies is that selective breeders have also increased the size of the males' tails to show off the brilliant color patterns, and these oversized tails have slowed their bearers to the point that they have difficulty catching up with the females whose attention they have attracted.

Selective breeders have also begun to develop colors in the females, which doesn't seem to affect the guppies' sex life, but it can be confusing to a human observer who expects to be able to distinguish the sexes by appearance. To tell a male guppy from a female guppy these days, you have to focus on the single fin on the guppy's underside—the one right in front of its tail, called the anal fin. In females, the anal fin is a normal fan shape, but in males it's pointed.

This pointed fin is crucial to the guppies' way of mating. Unlike goldfish, who mate externally—the female ejecting her eggs and the male showering them with his sperm—guppies mate internally, or almost internally. The female keeps her eggs inside her body, and the male does his best to shoot packets of his sperm into her genital opening by aiming his pointed fin, which serves as a channel, directly at it. The female can store whatever sperm she receives for several months, producing twenty to fifty young every four to six weeks.

The fertilized eggs develop inside the female, which creates the impression that she gives birth to live young. Actually, her young are just as egg-hatched as goldfish young. The eggs merely hatch inside, or just outside, her body. Most people who have owned guppies have watched a female grow fat and develop a dark spot on her enlarged abdomen where the embryos' dark eyes show through her skin. And perhaps they've even witnessed the birth of the young, who upon emergence swim as fast as they can for shelter because, within the confines of an aquarium, adult guppies cannibalize their young.

In the wild, cannibalism isn't a problem. The slow-moving tropical rivers where guppies evolved offer plenty of space and hiding places for the young, and the adults are too busy courting, mating, and escaping from predators to have time to chase their own offspring. The wild guppies' strategy is to produce great numbers of young—which accounts for their other common name, the "millions fish"—to maintain their own population amid the numerous predators who find guppies just the right size to eat. In turn, guppies prey on mosquito larvae and other small aquatic insects and browse on algae and organic debris, which helps to keep their river ecosystem in balance.

Guppies, who have taken so well to life indoors, got me thinking about survival, both ours and theirs. The kinds of organisms that can survive indoors are those who, like ourselves, are adaptable. We adaptable human beings moved indoors originally because it was conducive to our own species' survival, but apparently we weren't interested in surviving alone. We seem to have spent much of our indoor history searching for other species as adaptable as we are so that we would have some company indoors.

SEA HORSES

I encountered my first live sea horses at the house of a friend, who had advanced considerably beyond the ease of tap water in a goldfish bowl or even fresh water in an aquarium to the rigors of simulating the sea. She had several tanks of exotic ocean-dwelling fish, but from the moment I walked into her basement aquarium, I was captivated by her sea horses. They were the dwarf species, the kind you can send off for in the mail that grow to a size of about an inch and a half (3.75 cm), but they demonstrated all the attributes and behaviors of the bigger species you can observe at urban pet stores and at big marine aquariums.

I watched the dwarf sea horses drift gracefully around their gently bubbling tank, rising, descending, turning archly to the left and right, and propelling themselves short distances forward by vibrating an almost invisible fin halfway down their backs. When one reached its destination, it would twine its long tail around a stalk of imitation coral to anchor itself. As I observed these sea horses, I examined every detail of their unusual bodies to try to figure out what kind of animals they might be. But the best I could do was guess from their hard, segmented exteriors that they might be crustaceans of some kind—kin to lobsters, crabs, or shrimps. My friend laughed at my ignorance and explained that they, like her other charges, were ocean-dwelling fish.

If, in your imagination, you straighten out a sea horse's body, float it horizontally, convert its exterior armor to scales, shorten its snout, and exchange its long, tapering tail for a fin, you might be able to perceive its kinship to a fish. But looking at a sea horse this way reverses the evolutionary process. If you focus instead on how the sea horse differs from a fish, you can see how its differences have helped it find a special niche for itself among other inhabitants of the sea.

The sea horse's divergence began when the ancestral animal discovered that it could survive without speed. It began evolving away from the typical fish's speed-related form and toward body parts that suited it to a more sedentary existence. Because the sea horse found advantages in staying still, it evolved a long tail, which is prehensile like a monkey's, to anchor itself to pieces of aquatic vegetation and coral. It also shifted to a vertical posture to look more like its vertical supports and developed the ability to change colors to match its surroundings. When a sea horse does move, it propels itself not with its tail, like other fish, but with its small back fin. Even when it beats this little fin full speed, however, the sea horse proceeds so slowly that it looks as if it might have broken loose from a slow-motion merry-go-round.

Anchored in shallow coastal waters, or swimming slowly from one anchorage to another, the sea horse encounters plenty of food without having to chase it. Sea horses feed on small animals, such as brine shrimp, that are part of the ocean's plankton—the microscopic plants and animals that float together in great numbers near the ocean's surface. With two large eyes that move independently of one another, the sea horse notices all the minuscule activities in its vicinity, and when it spots a tiny animal floating by, it merely points its long snout in that direction and sucks in a meal. Because each meal is small, a sea horse spends much of its time eating.

Toward the end of winter, sea horses begin to look for mates. When a male and female meet, they dance around each other and intertwine tails in an elegant courtship ballet. Then, at the end of their dance, a strange thing happens: the female gives her eggs to the male. She deposits them in his brood pouch, which is located right where what looks like his round belly curves into his tail. The eggs are fertilized as they enter, and when the male's pouch is full, its walls swell into a spongy tissue that isolates each egg in a separate chamber. The eggs then hatch, and the developing embryos are nourished by egg yolk and the surrounding walls of the male's brood pouch.

After forty to fifty days—in the case of a dwarf sea horse, only ten days—the young are ready to be born. By this time the male's brood pouch has swollen to a considerable size, and he goes through his own version of labor. He twists his body, contracts his muscles, and even pushes his pouch against solid objects until the young begin to exit from a small opening toward the top of the pouch. In dwarf sea horses, the young usually emerge one at a time, but in the larger species they often emerge in spurts. While a dwarf male might produce only ten to thirty-five offspring, the bigger species produce two hundred to six hundred, all miniature replicas of their parents and completely independent as soon as they leave their father's pouch.

The dwarf sea horses you can order from marine supply houses are often pregnant males, who sometimes give birth in transit. Keeping these sea horses and their offspring as pets is somewhat more demanding than keeping goldfish or guppies, but if you are willing to commit yourself to simulating a marine environment and providing your dependents with a constant supply of live marine food, you can offer them a decent—if not exactly oceanic—life indoors.

II. AN INDOOR PLANT KINGDOM

AFRICAN VIOLETS

Long before I became a naturalist I had an unfortunate experience with houseplants. I was living in a one-room apartment in Washington, D.C., and wanted a room divider to create the illusion of separate spaces. It was winter, and the thought occurred to me that some greenery might both divide the room and cheer up my dark living quarters. So I went to the five and dime, bought a dozen or so different plants, hung some from the ceiling, and arranged the rest on a set of shelves below.

At first I was quite pleased with my organic "wall." But, having no understanding of plants, I left them all in their original positions and watered them all at the same time—profusely—whenever it occurred to me. Needless to say, some of my plants began to look straggly, while others became limp, and still others turned yellow.

By then it was early spring, and I decided that what the plants needed was some fresh spring air. I carted the whole bunch of them out onto my small balcony late one Saturday afternoon, went out for the evening, and completely forgot about them until the next morning. When I woke up, I discovered that a late frost had put an end to my houseplants' suffering, leaving me with such a whopping sense of guilt that I avoided houseplants as much as I could for over a decade. Then a friend gave me an African violet for Valentine's Day. I was pleased, of course, but I also felt a sinking feeling in my stomach: surely this healthy little plant with its beautiful purple flowers would die under my supervision.

During the intervening decade, I had moved to Vermont and developed a substantial interest in the natural world. Wildflowers, I had discovered, are guiltproof because they grow where they want to grow, in perfect harmony with the seasons, soil, sunlight, temperatures, and moisture that surround them. They are not dependent on human beings for anything except the courtesy of not stepping on them or picking them before they've had a chance to produce their seeds. I decided to commit myself to the African violet. I would do my best to simulate its natural living conditions and hope it would grow like a wildflower in my home.

The ancestral African violet came from the east coast of Africa, where it was discovered growing out of crevices on wooded cliffs near the city of Tanga in what is now Tanzania. At the time— 1892—this part of Africa belonged to Germany, and the German official who chanced upon these attractive wildflowers sent specimens home to his father in Germany. His father grew more from seeds and passed them around among his horticultural friends, one of whom referred to them as "violets" because the five-petaled, purple flowers resembled the flowers of the violet family. The plant is totally unrelated to the familiar wildflowers called violets, but the name "African violet" has stuck.

In its native habitat, before the residents of modern-day Tanzania changed the environment by clearing land for agriculture and cutting trees for firewood, the wild African violet grew in warm, tropical sunlight filtered by trees and shrubs. It was watered by moisture dripping down the cliffs and humidified by coastal mists and dews. The closest I've been able to come to these conditions in my home is a kitchen windowsill, which offers the pseudotropical warmth of my own indoor environment, sunlight filtered through the branches of a lilac, and the humidity generated by cooking three meals a day.

My African violet seems to be content with the basic arrangements. I try to meet its needs for water and nutrients simultaneously by putting some water in a bowl, adding a little commercial African violet food, and setting the African violet, pot and all, in the bowl. When the plant has "drunk" as much as it wants and is taking up no more water, I put it back on its saucer on my sunny windowsill.

The enthusiastic flowering of the African violet, which has made it such a popular houseplant, would in the wild promote pollination. The bright yellow structures at the center of each flower are the flower's male parts, packed with pollen. If you touch one, however, you won't set any pollen free because it's trapped inside a closed sac. The wild African violet depends on an insect—one expert says a thrip—to break into the pollen sac, eat some pollen, and crawl to another plant, carrying a bit of pollen on its body. The tall, hairlike structure at the center of the flower, rising above the male parts, is the female part, which needs a bit of pollen dusted onto its tip to produce seeds.

African violets rarely produce seeds indoors unless they are hand-pollinated, but they can reproduce themselves in other ways. As each plant grows, it creates little duplicates of itself, called offsets, which in the wild would become an expanding clump of genetically identical plants. If my African violet is satisfied enough with my support services to produce offsets in my kitchen, I will someday have to repot it, separating the offsets and potting them individually to become new houseplants.

If I get hooked on African violets and want still more duplicates, each leaf is also capable of producing a new plant. I will have to cut a healthy leaf from somewhere toward the middle of my plant, push its stalk through a covering of wax paper into a glass of water, and wait patiently for it to root and produce a new little plantlet.

If I do everything right and my African violet multiplies, my kitchen may indeed begin to resemble the cliffs of East Africa. Small purple flowers will bloom from every shelf and windowsill. Given the threats that wild African violets face in their native habitat today, I've convinced myself to rise above my guilt and do my best to perpetuate this one cultivated descendant that has found its way into my home.

HEART-LEAVED PHILODENDRON

Maybe I would have developed a happier relationship with house-plants if I had begun with just one instead of the assortment I tried. As it is, my traumatic experience makes it necessary for me to do most of my indoor botanizing in places other than my own home. But I can't say I have any difficulty finding specimens to observe. Houseplants have become so common that I see them everywhere I go—the grocery store, the Laundromat, the hardware store, even the bank. I study them at doctors' offices, restaurants, and beauty salons. Indeed, it's the beauty salon I go to to get my hair cut that has helped me reestablish a relationship with the heart-leaved philodendron, which was among the plants I killed.

Right above the mirror where I sit, the familiar dark-green, heart-shaped leaves tumble from a hanging basket. Several long stems wander horizontally across the top of the mirror, while another two or three drop down along one side. This hanging display, I've been told, was created with cuttings from a bigger heart-leaved philodendron that grows in the reception area. The parent plant lives in a sizable tub on the floor, with all its greenery growing in a dense column up a slab of bark. The beauty salon is in the process of being overrun by the offspring of this productive individual, but no one seems to mind.

To learn how a heart-leaved philodendron grows in the wild, you'll need to find a healthy specimen—like the one at my beauty salon—climbing up a slab of bark. In the tropical rain forest, where this climbing vine was discovered, it attaches itself to the bark of a tall tree with strong brown holdfasts called aerial roots. These roots are completely unrelated to the plant's feeding roots, which are anchored in the soil. They merely stick to moist bark without penetrating it or taking anything from the tree. If you look around at the back of the bark slab, you will be able to see these aerial roots more clearly against the sawed wood than you can against the bark. They radiate from each leaf joint like a spider's legs, attaching the long, weak stem of the vine quite firmly to its support. The heart-leaved philodendron's stategy is to scramble up a tall tree to gain access to sunlight.

In the tropical rain forest, a heart-leaved philodendron can grow without inhibitions. Frequent rains provide plenty of moisture, the sunlight that filters down through the canopy provides just the right amount of light, the soil provides a steady supply of nutrients, and tall trees allow the plant to grow as big as it wants. A wild heart-leaved philodendron grows much bigger than a houseplant, with leaves that reach one foot (30 cm) in length. According to its natural life cycle, an individual plant must achieve at least this size before it can flower.

In a North American home, limited by a pot, erratic watering, dim light, occasional feedings, and short supports, a heart-leaved philodendron never advances beyond its juvenile, nonflowering stage, which is characterized by small, 4- to 6-inch (10- to 15-cm) leaves. The ever-lengthening stem, especially if the distance between the leaves becomes excessive and the leaves themselves become smaller and smaller, is evidence of the plant's determination to cover distance if it can't gain size—in hopes of finding better growing conditions somewhere else.

This perpetually growing stem makes the heart-leaved philodendron easy to propagate. You can cut off a short length of the growing tip, set it in a glass of water until it roots, and then pot the new plant either in a hanging basket or in a tub with a slab or bark. Or you can cut a longer piece of stem, and in addition to rooting the tip in water, root several shorter sections that have at least one leaf by sinking them into moist perlite—a sterile, water-holding medium that professional plant propagators use. Roots will grow from just below the leaf joint into the perlite, while the eye—the little bud directly across from the leaf—will produce a new plant.

The ease with which a heart-leaved philodendron can be propagated is, in fact, what eventually made it such a popular houseplant. But it wasn't an instant hit. The ancestral plant was first discovered growing in the tropical rain forests of Jamaica by none other than Captain Bligh of *Mutiny on the Bounty* fame, but no one paid much attention to it when he delivered it to England in 1793. Perhaps the English horticulturists of that era were so overwhelmed by exotic plants arriving from all over the world that the unpretentious heart-leaved philodendron escaped their notice.

But during the next century, houseplants moved out of the greenhouses of collectors into the parlors and conservatories of wealthy plant lovers, and finally into the living rooms and kitchens of average folk. By the 1900s, everyone wanted houseplants, and the heart-leaved philodendron turned out to be perfectly suited to this new era of botanical democracy.

Because the heart-leaved philodendron was so easy to propagate, it could be sold at a reasonable price. Because it was tolerant, it could survive in the dim light and dry conditions of most homes, and because it would survive, it would bring pleasure to people who had no special botanical expertise. In 1936 an enterprising nurseryman in Florida saw the heart-leaved philodendron's potential and launched the modern era of houseplants when he decided to market heart-leaved philodendrons through five and dimes.

I've finally managed to develop an amicable relationship with the heart-leaved philodendrons that are proliferating at my beauty salon. I don't think I'll ever take one back into my own home, but it's been therapeutic for me to see them thriving elsewhere. Someday, the naturalist in me would like to deliver my apologies to the ancestral plants in their native rain forest, but in the meantime, I pay regular obeisance to their pampered progeny whenever I visit the beauty salon.

WANDERING JEW

I am confronted by another of my victims almost every time I turn around: the wandering Jew. This trailing vine is one of the most indestructible houseplants known to modern horticulture, but I killed it, too. Whereas I thought my wandering Jew's long, straggly stems meant the plant needed some fresh spring air to perk it up, they actually meant I should have cut them back a long time ago. Instead of putting the plant out on my balcony that ill-fated day, I should have cut off most of the stem tips, repotted them in fresh potting soil, and thrown the old plant, which was dying, away. But I didn't know anything about wandering Jews back then, so I was at the mercy of my houseplant's rampant growth.

Wandering Jews are designed to continue growing almost no matter what happens to them. Like heart-leaved philodendrons, they are weak-stemmed vines adapted to life in the tropical rain forest, but they have evolved a completely different strategy for survival. Instead of climbing up trees to find sunlight in the canopy, they clamber over the forest floor, gathering their sunlight from the scattered rays that filter through the canopy.

Because the plant's growing tip is always moving farther away from its original roots, the wandering Jew would be hopelessly vulnerable if it didn't develop some additional roots along its length. If you examine the stem of a wandering Jew, you will notice that each leaf is attached at a slightly swollen joint. If one of these joints, which are called nodes, touches soil, it sinks roots, providing the growing tip with a closer source of moisture and nutrients. If something happens to the original roots, or if they just die of old age, the growing tip can still rely on whatever new roots have grown along the way.

If something happens to the growing tip, on the other hand, the remaining stem responds by producing branches. The branches emerge from buds hidden in the leaf axils—the angles where the leaves join the stem. These axillary buds are dormant on a healthily growing stem, but they are ready to produce new growing tips should the main tip disappear. One way or another, the wandering Jew deals with distances and disasters, perpetuating its genes as it rambles across the rain forest floor.

Periodically it also offers to share its genes with other members of its species by flowering. The wandering Jew flowers more readily than the heart-leaved philodendron and often produces its small white blossoms indoors. Each is three-petalled with six yellow-tipped male parts and numerous white hairs at the center, but you need a hand lens to observe these details because the whole structure is only half an inch (1.25 cm) across. Because the flowers are so small and each one lasts only a day or so, the wandering Jew is thought of less as a flowering plant, like an African violet, than as a foliage plant, like a heart-leaved philodendron.

As a houseplant, the wandering Jew is usually grown in a hanging basket. That way its attractive foliage, which was shiny green in the species I owned and is variegated or striped with purple in other species, can cascade over the edges of the pot instead of sprawling across a tabletop or shelf. Because the stems are growing in the air, however, they don't have an opportunity to root at their nodes and will just get longer and stragglier if you yourself don't encourage them to exercise some of their other options.

If you remove just the smallest pairs of leaves at the growing tips with your thumb and fingernail—a process called pinching—the long stems will branch to produce a fuller, bushier display. If the bases of the stems seem to be getting old and dying, you can create a new display, or several, by cutting off three or four inches (7.5–10 cm) of each stem tip, breaking off the lower leaves, and sinking these cuttings directly into a pot full of moist potting soil. These tip cuttings will root and lengthen, and within a month or so, they, too, will be cascading over the edges of their pot.

The wandering Jew's ability to keep growing in almost any circumstances is legendary. A researcher in California once put some wandering Jew cuttings to tests about as severe as any houseplant is likely to encounter. First he just left the cuttings lying on his work table, where their only resources were their own physical substance and whatever moisture they could absorb from the atmosphere. These cuttings lived for almost two years and even grew in length during that time. The bottom leaves died, and the new leaves were exceptionally small with exceptionally short sections of stem between them, but the cuttings survived.

Then, after two years on the researcher's work table, some of the cuttings were put into potting soil, where they rooted and began to grow like typical wandering Jews. Others were submitted to another environmental extreme. Having survived a two-year drought, they were submerged in running water to see if they could survive a one-month drowning. They grew less under water than they had in dry air, but they managed to stay alive.

The one thing this resourceful plant cannot tolerate, as I so sorely learned, is frost. Its moist and sensitive tissues, which serve it so well in the tropical rain forest, have no defense against freezing. A wandering Jew can live quite comfortably in the pseudotropical environments we human beings create for ourselves in northern climates, but in becoming a houseplant it has wandered too far from its native habitat to fall back on its own devices should it find itself, unseasonably, outdoors.

JADE PLANT

It's embarrassing enough to me as a naturalist to have killed two almost indestructible vines, but I also annihilated a succulent. Succulents are tough plants with special moisture-conserving adaptations that enable them to survive in some of the most severe environments on earth. They grow in parched deserts and windy, stony wastelands, where it rains only a few times a year. But, like tropical vines, warm-climate succulents are vulnerable to frost. The particular succulent I did in is a native of South Africa—the jade plant.

A domesticated jade plant looks like a miniature tree. Its thick, fleshy stem turns brown with age, and the new growth branches freely, creating old angles and symmetries, until the whole plant begins to resemble a gnarled and ancient bonsai. A bonsai is a tree that has been dwarfed, pruned, and shaped according to methods devised by the Japanese so that it can be displayed in a small pot. But whereas creating a traditional Japanese bonsai entails radical surgery to roots and branches, a jade plant just grows that way.

In its native South Africa, a jade plant can reach a height of 10 feet (3 m) and look more like a real tree than a bonsai. It flowers profusely during the dry South African winter, bearing little pyramids of pinkish-white blossoms at the tips of its branches. As a houseplant, however, a jade plant rarely grows taller than 3 or 4 feet (.9 or 1.2 m) and rarely flowers. Some large indoor specimens that are twelve years old or older can sometimes be induced to flower by providing them with lots of water and sunshine during the summer and then withholding water during the winter. But smaller specimens, like the young thing I bought at the five and dime, will merely produce more leaves.

These leaves give the jade plant its common name—they look and feel like polished jade—and their unusual thickness explains the jade plant's ability to survive in dry environments. Within the group of plants called succulents, the jade plant is known as a leaf succulent (as opposed to a stem succulent) because it stores its water in its thickened leaves. During the sporadic summer rains, a wild jade plant absorbs as much water as its leaves can hold and then uses this stored moisture to support its activities during the rest of the year.

In addition to storing water, a jade plant's leaves also conserve it. First, the thick, waxy cuticle, which gives the leaves their shine, prevents moisture from evaporating into the atmosphere. But the jade plant has a second, somewhat unusual adaptation for protecting its limited supply of water. Plants must photosynthesize during the day, when the sun is shining, to provide the energy for this food-making process. And most plants open their stomata, or leaf pores, while they're photosynthesizing to take in the carbon dioxide that mixes with water to become food. But open stomata mean escaping moisture, especially if the atmosphere is hot, dry, or windy.

The jade plant cleverly opens its stomata at night, thereby sacrificing less moisture to the atmosphere. The carbon dioxide that comes in through the stomata mixes with leaf fluids to become an acid, which the leaves can store until the sun shines. Then, during daylight hours, a jade plant photosynthesizes along with all the other plants by making use of its stored carbon dioxide. This water-conserving approach to photosynthesis is called *Crassulacean acid metabolism*, or CAM, for the Crassulaceae family, to which the jade plant belongs. But it is not limited to this family; several other succulents practice CAM, too.

If you look closely at how a jade plant grows, you will see another feature of the multitalented leaves. The plant's new growth comes right out of the bases of the old leaves. As long as the leaves stay attached to the stem, this growth pattern merely leads to the angles and symmetries of the whole plant, but if a leaf falls off or if you break one off, it can develop into a new jade plant.

First the base of the fallen or broken leaf must heal itself with wound tissue. Then the isolated leaf sets to developing roots and a new plantlet. Unlike the African violet leaf, which needs some pampering to produce its new plantlet, a jade plant leaf will respond to the surface of its parent's potting soil if it happens to fall there. Or it will sprout in a dish of sand. You can also propagate jade plants from stem cuttings by sinking 3-inch (7.5-cm) stem tips into moist but well-drained sand. In either instance, you will need to pot your new jade plant in a sandy potting mix suitable for succulents once it has grown its roots.

The way a jade plant lives in the wild, with rains that usually arrive in brief but heavy downpours and long dry spells in between, it probably stood a better chance than the other plants I bought of surviving under my regime. Most houseplant books even recommend allowing a succulent's soil to dry out between waterings. If only I hadn't frozen my young jade plant, I might now have more confidence in houseplants thanks to a rugged little tree.

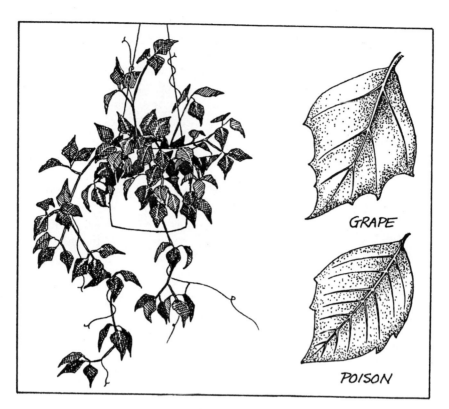

GRAPE

POISON

GRAPE IVY

Not every houseplant makes me feel guilty. Several common species weren't on display the day I shopped for my room divider and therefore escaped the ensuing disaster. These I can look at without remorse. Once I even took in an especially vigorous-looking individual as a foster plant, promising its former owner, who was moving, that I would find it a good home. This temporary resident, which lived with me for five years and survived three of my own moves, was a grape ivy.

First it lived in my dark city apartment, where it mostly accumulated dust. Then it followed me to my house in Maryland, where it got more light but about the same amount of attention. And finally it moved with me to Vermont, where it encountered and adapted to the rigors of country living, wood heat and all. This amazing plant was no beauty by the time I found it the good home I had promised, but at least it was still alive.

When I chanced upon a much healthier, better-tended specimen of this popular houseplant several years later, I had an odd experience. When I first saw it, I thought the garden center I was shopping at had potted up some poison ivy and hung it from the ceiling as a cautionary display. I was studying the display—with my hands safely in my pockets—when I suddenly noticed that the plant bore long tendrils on some of its stems. Poison ivy climbs by means of short aerial roots, somewhat like those of the heart-leaved philodendron, so the long tendrils seemed out of place.

I looked the whole plant over more closely and discovered that it wasn't poison ivy at all. When I focused on the shiny, green, three-part leaves, which had looked enough like poison ivy to deceive me, I saw that they were actually quite different. Whereas poison ivy's three leaflets are arranged with the two side leaflets growing close to the stem and the middle leaflet growing on a longer stalk, all three of the leaflets I was looking at grew on their own little stalks. And the leaflets themselves were more angular than poison ivy leaflets. The long, branched tendrils began to make me think more of grapevines than poison ivy. Then, when I reached up to turn the plant and look at it from another angle, I recognized the growth pattern of my old survivor, the grape ivy.

In the time since I had given the grape ivy away, I had learned to identify and avoid poison ivy, to identify and eat wild grapes, and to pay close attention to how wild plants grow. So I looked at the grape ivy with new eyes. I saw not the familiar houseplant but a wild green vine that grows naturally somewhere on this earth. I wondered where grape ivy was growing when it was discovered and what it looked like growing there.

Grape ivy hails from the misty mountain cloud forests of the West Indies and northern South America. In its native habitat, it scrambles over the forest floor and climbs partway up trees by attaching itself to twigs, branches, and other vines with its long tendrils. It can survive in the limited light that filters through the forest canopy, but it responds to brighter light by growing faster. It can also tolerate both the heat of the lower elevations, where it grows more rapidly, and the cold of the higher elevations, where it grows more slowly. Basically, it responds to the various and variable conditions it encounters by speeding up or slowing down over a wide range of tolerances.

This wide range of tolerances—including a tolerance of human neglect—has made grape ivy a highly successful houseplant. It is most often grown as a hanging plant, which shows off its evergreen leaflets and trailing stems but doesn't make much use of its tendrils. If you want to see how the tendrils work, you can set your grape ivy on a shelf or table, or even on the floor, and offer it some strings or a trellis to climb up. When its sensitive tendrils encounter something to hang onto, they curl around it, enabling the weak stem to climb vertically.

In the wild, a grape ivy reproduces itself by flowering and producing seeds. As a houseplant, however, it will never grow big enough to flower. It must be propagated by cuttings, which can be rooted in water or in other rooting mediums, such as perlite. Like many of the other clamberers and climbers that have become houseplants, grape ivy is capable of growing roots from each leaf node, which in the wild allows a broken stem to become a whole new plant.

In thinking about grape ivy, both the healthy, thriving specimen I saw on display at the garden center and the straggling old survivor I finally placed in a good home, I was struck by how the outdoors has brought me back indoors in a way that the indoors never took me outdoors. When I tried to keep houseplants, I never really looked at them or learned to understand them as individuals or even as types. It took wild plants, with their bewildering array of differences, to start me paying close attention to identifying characteristics and to just how each plant lives, grows, and perpetuates itself in response to its environment. Knowing what I know now, I have a new respect for my persistent old grape ivy. It was not luck that enabled this plant to survive my various and variable life-style. Rather, it had exercised some of the extreme tolerances that make this species a success in the misty mountains where it grows.

ENGLISH IVY

Another common houseplant I chanced not to buy—and hence can view without remorse—is the hardy favorite called English ivy. Ivy is not a technical word describing botanical features shared by grape and English types. Rather, it is a popular word that English speakers apply somewhat randomly to various plants that creep and climb.

Whereas most of our common houseplants come from the tropics, English ivy comes from the temperate zone. It arrived in North America with the English settlers, who knew it as a beloved wild vine that clambered over rocks and stumps or climbed right up the trunks of trees. From these colonial and later plantings, English ivy has escaped into the wild and now grows as a naturalized outdoor plant as far north as New York and Massachusetts.

Because English ivy is a temperate plant that grows outside our houses as well as inside, it offers an indoor naturalist the opportunity to compare a houseplant to its own wild version. In the wild, English ivy grows both horizontally and vertically. You will see it growing in beds and also climbing trees and buildings. You can observe both these growth habits indoors by planting one English ivy in a hanging pot, with its would-be horizontal stems tumbling over the sides, and another in a standing pot with a slab of bark for the vine to ascend. At first you might need to tie the English ivy's stem to its support, but eventually the vine itself will take hold.

Whereas grape ivy climbs by means of clasping tendrils, English ivy, like heart-leaved philodendron, uses aerial roots. Your hanging plants won't produce these aerial roots, but if you examine one of its stems, you will see some little dots just below each leaf, which indicate places where they could grow. Interestingly enough, only young English ivy plants produce these aerial roots because they are part of the plant's youthful strategy for reaching the sun. Once the growing tip of the vine has reached the top of its support, which might take ten to fifteen years in the wild, the new growth changes from the juvenile, vine-lengthening phase to a mature, flowering phase.

Indoors, English ivy lacks the space and sunlight to mature. Just reaching the top of a slab of bark is not enough to motivate a life change. Even climbing to the top of a window frame and wandering around close to the ceiling isn't enough to deceive the English ivy because what it is looking for as the vine lengthens is more sunlight. So you'll have to look outdoors to see mature English ivy.

The mature plant looks more like a shrub than a vine because its stems and branches are shorter with more leaves crowded closer together. Whereas the juvenile leaves—the familiar three- to five-lobed, pointed ones you see on houseplants and the lower portions of wild plants—are adapted to making do with the bit of sunlight they receive through the leafy canopy of a tree, the adult leaves are adapted to life in the sun. They are rounder and thicker, and they make more efficient use of the full sunlight they enjoy.

The additional sunlight and the mature leaves' more efficient use of it provide the wild English ivy with the extra energy it needs to flower and produce its seeds. In the fall, the mature plants produce clusters of sweet-smelling greenish flowers at the tips of their branches. The flowers are pollinated by visiting insects and develop into small, dark fruits that fall to the ground to start the English ivy's long life cycle all over again.

Even though any English ivy you keep as a houseplant is

doomed to eternal youth, it can still reproduce itself—but not by means of flowers and seeds. If you cut off a 3- to 4-inch (7.5- to 10-cm) length of stem tip, which, incidentally, will encourage that stem to grow branches, and put it in a glass of water, it will produce new roots. You can either leave the rooted cutting in the water as the beginning of a water garden—to which you can add cuttings of heart-leaved philodendron, wandering Jew, and grape ivy—or you can plant it in a new pot.

If you're interested in other indoor experiments, you can drape a whole section of stem that's still attached to your hanging plant across the surface of some moist perlite and watch it root at the leaf nodes. Or you can cut a length of stem into short sections, each bearing two or three leaves, remove the lowest leaf, and stand each cutting upright in the perlite. Each cutting will root and when potted will produce a whole new plant.

If you'd like to duplicate an experiment Darwin tried, you can tempt an English ivy to climb a piece of glass. Darwin's plant adhered to the glass and "secreted a little yellowish matter," but my plant refused to cooperate. Even though I kept the stem tied to a piece of glass for several months, it never took hold.

English ivy is such a persistent and durable plant that it gives even the most inexperienced naturalist a chance to experiment and observe. And because you can observe it both indoors and out, you can compare your houseplant to its wild form. This temperate vine may not seem very glamorous or exotic, but for those who share my fear of houseplants, English ivy offers demonstrable proof that the species will survive whatever happens to the individuals who happen to live indoors.

SPIDER PLANT

I seem to see spider plants growing in more places than any other houseplant. The individual I have come to know best lives at my Laundromat, where it survives quite nicely in the warm, moist environment created by busy washing machines and dryers. Once I spent an entire wash cycle examining this spider plant, whose numerous little offspring dangled above the churning machine as if they might drop right in.

The spider plant is well named. Its slender, arching leaves look like the legs of a spider, and the miniature plantlets that grow from long stems, called runners, look like spiderlings hanging from strands of silk. These spiderlings are such intriguing little duplicates of the parent plant that I had to look one over upside and down.

Growing upward was a tuft of five short leaves that looked like blades of grass with white edges. At the base of the tuft was a cluster of fat nubbins that looked like dried roots. These nubbins are aerial roots, like the heart-leaved philodendron's holdfasts, but they looked as if they might grow into regular underground roots if I offered them some soil. I was curious enough that I asked the owner of the laundromat if I could have a few spiderlings for some home experiments, and he was only too happy to trim a handful off his prolific plant for me.

I decided to suspend one of the spiderlings over a glass of water with just its aerial roots submerged so I could watch what would happen. The others I planted in small pots filled with potting soil. All my spiderlings stayed green and some even produced new leaves, but nothing else observable happened. After six weeks, I dug up one of the spiderlings I had planted in potting soil to see if its roots looked any different from the roots that had been submerged in water. I was surprised to discover several thick, white structures that looked like white icicle radishes projecting far beyond the nubbins, which, like those in the water, hadn't changed.

Some of these thick white roots bore long, thin hairlike roots, which made the spiderling difficult to extricate from the soil. These root hairs are the spider plant's working roots. They absorb water and nutrients from the soil to support the plant's growth, while the thick white roots are special water-storage organs to help the spider plant cope with dry weather. In the part of South Africa where spider plants grow wild, which isn't too far from where jade plants grow wild, the rainfall is seasonal, falling mostly during the summer, so the water-storing roots are necessary to keep the plants alive during the dry winter.

As a houseplant, the spider plant is usually displayed, like grape ivy, hanging from the ceiling. This practice creates the impression that a spider plant grows high up in trees, but in its native South Africa, it lives on the ground. It grows like a big clump of grass, sending out its runners to establish duplicates of itself nearby. At first each runner grows upward from the center of the leaves, but as the weight of the spiderling—or sometimes two or three spiderlings—increases, it arches and bends down to touch the ground. Once the plantlet finds soil, it develops its own feeding and water-storing roots, so it no longer needs nutrients and water from the parent plant. The connecting runner degenerates, and the new plant—a genetically identical replica of its parent—becomes completely independent. Eventually it grows big enough to send out plantlets of its own.

Of all the common houseplants, the spider plant is probably

the easiest to propagate. You don't have to nurture stem tips or leaf cuttings in a glass of water or a rooting medium. All you need to do is offer a spiderling its own little pot, and you have an instant gift to give a friend—which is how the spider plant got popularized as a houseplant in the first place. In 1828, the German poet and naturalist Johann Wolfgang von Goethe received a spider plant for his botanical studies, and he was so impressed by the spiderlings that he proceeded to give them away to all his friends. More than one hundred fifty years later, we are perhaps beginning to see the full impact of Goethe's enthusiasm.

Spider plants have become so exceedingly popular in large part because they can endure almost anything human beings do to them. They can live in lobbies, waiting rooms, libraries, dimly lighted bathrooms, children's bedrooms, and bustling Laundromats. They might also make good additions to laboratories, because recent studies have revealed that they can lower the level of formaldehyde in the atmosphere.

The only indoor condition that some spider plants have trouble with is our long, artificially illuminated days. If a spider plant is located where it gets both natural daylight and several extra hours of electric light at night, it may just continue to grow leaves without ever putting out the runners that bear the spiderlings and, sometimes, small white flowers. To produce runners, some spider plants need at least three consecutive weeks of eight-hour days, a schedule that would be easy enough for most of them to come by during the winter if they didn't have to contend with electric lights.

The Laundromat's spider plant, which seems to do quite well on twelve-hour days, looked so healthy that I hadn't hesitated to ask the owner for spiderlings. But after they had rooted, I realized I had a problem on my hands. If my spiderlings prospered, I'd soon be responsible for a lot more spiderlings. And the thought of a house full of spider plants, all of them reproducing at the geometric rate that characterizes this species, helped me decide to return the spiderlings to the Laundromat. I transplanted all of them into one hanging planter and presented them to the owner. He wasn't exactly pleased to see his spiderlings back, but because I'm one of his valued customers, he thanked me and set about looking for another hook to hang yet another houseplant on.

BOSTON FERN

A friend who heard I was interested in houseplants invited me over to meet her fern. I expected to be introduced to some frondy greenery off in a dark corner and felt a bit silly as I drove to her house. The way she had been raving about this particular fern, I was prepared to act impressed, but I wasn't prepared for what I saw. Hanging from her living room ceiling in a huge copper planter was the biggest fern I'd ever seen. She led me over to the planter and said, "I'd like you to meet my Boston fern. I call it Sir."

Sir exaggerated almost every feature of the Boston fern. This popular houseplant is famous for its lush appearance and its long, gracefully arching fronds, but Sir had lusher, longer fronds than even the biggest of Boston ferns I've seen in greenhouses. Some of Sir's newer, shorter fronds filled the pot with a thick mass of upright vegetation, while older fronds arched toward the ceiling, and the oldest hung downward almost to the floor. I could have climbed inside Sir's cascading foliage and disappeared completely, but my friend teased me that pets and neighborhood children had been lost that way.

I became convinced that Sir was a mutant, a chance variation of the Boston fern that happened to carry the genetic potential for gigantic growth. My friend had encouraged this potential by offering her growing fern bigger pots, providing it with ample humidity, water, and nutrients, and displaying it in an optimal spot in her bright and airy living room. I'm sure that she also talked to it, becoming increasingly humble as the fern began to reveal its potential.

The idea of mutation is not far-fetched with Boston ferns. The first Boston fern was itself a mutant. During the 1890s, North America was experiencing a belated version of the fern craze that had swept England during Queen Victoria's reign (1837–1901). Wild Florida sword ferns—handsome, durable ferns with stiff, erect fronds that look like swords—were being shipped north in considerable numbers to meet the sudden demand for "a fern in every parlor," when among them a Boston seller noticed an individual that looked as if it belonged to a different species. Its fronds were broader and more graceful, arching from the pot rather than standing stiffly erect.

The year was 1894, and it took another year or so for the fern experts to decide whether this fern represented a new species or was merely a mutant of the sword fern. They finally determined that it was a mutant, whereupon fern growers began to notice that this new fern, named the Boston fern for the city in which it was first noticed, was genetically unstable. It was producing more mutants that were even bushier, lacier, and more elegant than the original Boston fern. By 1920, when the North American fern craze ended, seventy-five different mutations of the Boston fern had been described.

Mutant ferns are different from the hybridized flowers that horticulturists are so fond of developing. Closely related species of flowering plants can be intentionally hybridized—that is, interbred—by hand-delivering ripe pollen from one species to the open flower of the other. The seeds that develop from this deliberate interbreeding carry genes from both species, and the plants that grow from the seeds sometimes express new traits.

But ferns don't flower, produce transportable pollen, or grow from seeds. They are more primitive plants that reproduce by means of spores—single-celled reproductive units that must grow into intermediate plants before they produce new ferns. The early stages of a fern's reproductive cycle are microscopic and were not even understood until the Victorian fern craze inspired growers to pay close attention to such matters. And even after the reproductive cycle was clearly understood, the best fern growers could do was collect microscopic spores from the wild, germinate great numbers of them in their greenhouses, and wait for the ferns to appear. Ferns do sometimes hybridize if spores from closely related species happen to germinate close to each other, but it's a much less manipulable process than the hybridization of flowers. Thus, the Boston fern, with its tendency to produce mutant offspring, offered fern growers the opportunity to select and promote ferns with new traits without having to worry about hybridization.

Any mutation that appealed to a fern grower was fairly easy to reproduce in quantity because the Boston fern, when it's not mutating, can be encouraged to create numerous genetically identical duplicates of itself. The Boston fern's rhizome—a thick underground stem that supports the fern's aboveground growth from year to year—produces more than one plant. This rhizome can be dug up and cut into sections, and each section planted in its own pot becomes a separate fern.

The long leafless stems that appear among the Boston fern's fronds are runners, similar in function to the spider plant's runners. These, too, can produce genetically identical Boston ferns. But instead of producing new plants in midair as the spider plant does, the Boston fern's runners wait until they come into contact with soil. You can create as many duplicates as you want of a favorite Boston fern by offering its runners their own pots.

I'm not sure how many duplicates Sir would be capable of producing, but my friend is sure she doesn't want to take the risk of finding out. The way Sir looks right now, all this gigantic fern would need is a little such encouragement, and its offspring would gladly take over her entire house.

NARCISSUS

Winter can be a restless time for a northern naturalist. Snowflakes and animal tracks are interesting enough, and I enjoy watching the birds at my feeder, but, still, for most outdoor organisms winter means dormancy. One Christmas a friend gave me three paperwhite narcissus bulbs to help me through the long weeks until spring.

I wasn't sure exactly what to do with them because I'd never grown bulbs before, so for several weeks I left them in their Christmas wrapping. I almost forgot about them until one restless weekend, when I was wandering around the house in search of something to do. I happened to notice the package of bulbs, and for the first time I really looked at them.

They looked like onions, only with a thicker, browner skin. This outer skin is called a tunic, and bulbs that have one, such as onions, narcissus, daffodils, and tulips, are called tunicate bulbs. These particular tunicate bulbs didn't look very alive to me, and I was afraid I might have killed them by ignoring them for so long. But the plant kingdom is persistent, and it would take more than my inattention to destroy the life that waits patiently within a dormant bulb.

A bulb represents a survival strategy based on an annual period of inactivity, usually occasioned by a drought. It doesn't matter to the bulb whether it is buried in its native soil, stored at a commercial greenhouse, or left unattended in my cold Vermont farmhouse. As long as the environment is dry, its internal messages say, "Wait."

But a waiting bulb has everything it needs to start growing as soon as those messages change. At its center is a fully formed bud, which is the beginning of new leaves and flowers. Enough food to support the plant's early growth is stored in surrounding scales—the thickened bases of modified leaves. You can observe a bulb's basic structure by cutting into a plain old onion, which happens to be a bulb we cultivate for food, as opposed to the narcissus, which is poisonous, but which we cultivate for its scent and the beauty of its flowers. In addition to the papery tunic, the thick, concentric food-storage scales, and the bud, the onion shows the solid core of stem tissue that holds a bulb together and later will produce roots. This flat structure at the base of the onion is called the basal plate.

Because I want to eat my onions, I store them in a dry place to keep them from sprouting. With my narcissus bulbs, however, I wanted the opposite effect. I wanted to force them—that is, coax them into activity outside their natural growing season. It's fairly easy, I discovered, to simulate the conditions narcissus bulbs are waiting for. By the time the bulbs arrived at my house, they had already experienced the rest period they needed—the equivalent of a hot dry summer in their native Mediterranean habitat—before they would be able to grow again. This summer rest period, which is similar to winter dormancy in a colder environment, prevents bulbs from sprouting when there would be no water, which would, of course, mean death to a new young plant.

Once this requirement for rest has been met, however, all the bulb needs to become active is moisture, sunlight, and warmth. Flower books give detailed instructions about various potting mediums, room temperatures, and amounts of light and water, but I just stuck my bulbs into a pot of moist pebbles, set them on a windowsill in my living room, and waited to see what would happen.

I was delighted when my dry little narcissus bulbs produced

green shoots, assuring me that they were still alive. After the leaves had been growing for a while, each bulb also grew a single flower stalk. Exactly five weeks after my decision to activate the bulbs, the first delicate white flowers scented my living room, while yet another snow had just fallen outdoors.

The flowering of my narcissus bulbs represented the beginning of the end for them, which is the price such sensitive, warm-climate bulbs must pay for having found their way—via the horticulturists who have cultivated and hybridized them—from their native Mediterranean habitat to a subarctic environment like Vermont. In the wild, flowering is only one brief stage in the narcissus bulb's annual cycle. After the flowers have died, the leaves remain busy converting sunlight into the substance of next year's bulb. But in a pot of pebbles in a human living room, a narcissus merely consumes its stored energy and dies. Even if I had nurtured my bulbs in nutrient-rich potting soil and transplanted them outdoors in the spring, they would not have survived the outdoor seasons this far north.

I wouldn't feel comfortable tricking narcissus bulbs into blooming in my living room every year, only to throw them away at the end of their performance. But it was such a pleasure to have something alive and growing to take my mind off winter that some year I might experiment with the narcissus' hardier relative, the daffodil.

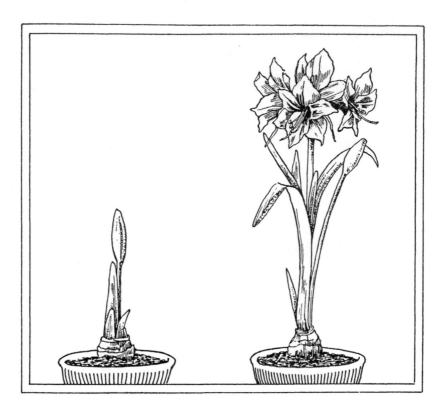

AMARYLLIS

Among my Christmas presents the next year was an amaryllis bulb. It was already buried to just the right depth in just the right kind of potting soil, so that all I had to do was add water. Buoyed by the success I had enjoyed with my narcissus bulbs, I duly added the water and started a process that still amazes me in retrospect.

First, the bulb, which was much bigger than the narcissus bulbs, loosened a bit and emitted what looked like green tongues. Then one of the green tongues turned into a cylindrical, rapidly growing shoot that gained at least an inch (2.5 cm) a day. When it reached 12 inches (30 cm) I was impressed, but when it hit 20 inches (50 cm) I began to worry about what might happen next. Meanwhile, the top few inches had been swelling into a bud that looked increasingly explosive. Somewhere between 20 and 24 inches (50 and 60 cm), the bud opened slightly to present one bright salmon-colored flower—then another and another and another, until four huge trumpet-shaped blossoms bowed gracefully from the top of the first flower stalk.

While the first four flowers were opening a little wider each day, a second flower stalk initiated its rapid ascent, and between the two cylindrical flower stalks, four flat, narrow leaves appeared. This single amaryllis bulb was offering a more dazzling display of energy than many plants show in a lifetime, and it wasn't finished yet.

I watched the dramatic changes from day to day, keeping notes the same way I do when I'm observing the humbler wild flowers that grow outdoors. As I watched, I couldn't help wondering where such a huge flower grew in the wild, what its natural cycle was, and what its desires and intentions would be if it weren't isolated in my living room.

The wild ancestors and relatives of my potted amaryllis, which is a hybrid of several species, grow in South America. But they are not residents of the tropical rain forest as so many other houseplants are. Rather, they grow in the Andes Mountains of Peru, where they must endure an annual dry season. Like a narcissus, an amaryllis survives the dry season by storing its energy in a bulb, but the bigger amaryllis bulb is much hardier. It can't be grown outdoors this far north, but it can be nurtured outdoors during the summer and will flower again indoors the next winter.

An amaryllis can indeed be perpetuated for many years if you are willing to commit the time and effort to simulating its native seasons. After you've tricked the bulb into blooming indoors midwinter, you'll need to water and feed it until it's ready to become dormant again. The flower stalks will die down after the flowers go by, but the leaves will stay green. Until May or June you will have to keep the plant indoors to protect it from frost, but about the time you set out your tomatoes, you can plant your amaryllis—pot and all—in a partially shaded place in your yard.

The amaryllis does not like to have its roots disturbed, so you should keep it in its pot when you shift it around. And even when the pot is buried outdoors, you should continue watering and feeding the plant to provide the large leaves with the moisture and nutrients they need to manufacture lots of food and store it in the growing bulb at their base. Before the fall frost, the leaves will turn yellow, signaling that it's time for you to simulate the Andean dry season by digging the potted amaryllis and storing it in a dry place—

your basement perhaps—for at least a month. By midwinter, the bulb will be ready to produce its amazing flowers again when you offer it some water.

If my amaryllis had been growing according to its natural rhythms in Peru, each flower would have been doing its best to attract an insect pollinator. Certainly the size and color of a wide-open amaryllis flower would catch the attention of passing insects, and the relative positions of the flower's male and female parts are designed to take precise advantage of their visits. The six male parts mature before the female part, offering their yellow pollen to the underside of the first several long-tongued insects who pause to feed. A few days later, the single female part curls upward to assume a position near the male parts. Its tip is then aligned to receive pollen from the underside of the next long-tongued insect who visits, which by virtue of the timing bears ripe pollen from another flower, assuring cross-pollination.

Because the amaryllis is so big, its flower parts and positions are easily observable. Even without a magnifying glass or slow-motion photography, you can see exactly what the flower is doing. The only element that's missing is the South American insect pollinator, who doesn't come with the hybridized, packaged, amaryllis bulbs sold at North American garden centers.

Although I am at best a reluctant keeper of houseplants, when friends set me up, as they have with an African violet, a temporary grape ivy, and twice with bulbs, I do my best to follow through. With the amaryllis, I fully intended to nurture the bulb through its annual cycle and invite it to bloom again the next winter, but the tall, top-heavy flowers in the lightweight plastic gift pot spelled doom. About the time the plant was at its most magnificent, it fell off my windowsill onto the floor, breaking its leaves and flower stalks in the crash. I buried the remains of this South American beauty in the cold New England earth.

ALOE VERA

When I first moved to Vermont, I had a housemate who swore by herbal remedies. She was always mixing up poultices, brewing herbal teas, and storing pulverized plant mashes in the refrigerator. The only one of her remedies I actually saw work and therefore came to believe in myself was the pulp of the aloe vera leaf. I now keep a potted aloe vera in my kitchen, but I think of it less as a houseplant than as a first aid kit. Whenever I burn, nick, or scratch myself, I cut off a bit of leaf and spread some cool, soothing moisture from the thick gelatinous interior directly onto the wound. The pain disappears and the injury heals without my having to think about it again. I would never use aloe vera on a deep cut or a serious burn because I've heard tales of healed skin hiding a deeper injury, but I depend on it to treat the minor assaults I seem inclined to inflict upon my outermost self.

Doctors and chemists have had some difficulty analyzing exactly what it is in the aloe vera pulp, which they call the gel, that takes the pain out of a skin injury and helps the tissue to heal so efficiently, but some of them are willing to concede that it works. Research has shown that aloe vera gel contains a compound that resembles aspirin in its ability to relieve pain and inflammation, and that it also contains substances that promote the growth of new skin cells. Furthermore, it seems to possess some antibacterial action, because an ointment containing an extract of aloe vera has performed as well as a plain ointment combined with penicillin. And doctors and chemists are not the only ones who are experimenting with aloe vera. Veterinarians have also used it to treat skin conditions in dogs, cats, and horses with some success.

Aside from its potential as household first aid, aloe vera is attractive enough to be kept as a decorative plant. Its thick, stiff leaves grow in a spiral from a short stem, with fresh new leaves always emerging from the center. The white spots that mark the leaves of an immature aloe vera create an appealing variegation, and the small spines that grow along the edges of the young leaves are not sharp enough to hurt fingers.

Wild aloe vera evolved on dry windy islands off the coast of West Africa, where it had to adapt to long periods of drought broken by sparse winter rains. Like the jade plant, aloe vera is a leaf succulent that practices *Crassulacean acid metabolism* (CAM). Its leaves swell with moisture during the rainy season, retaining it for use during long dry spells, and open their stomata at night, storing carbon dioxide for use during the heat of the day.

Many low-growing leaf succulents have spines like the aloe vera's to protect their supply of moisture from thirsty animals. Aloe vera leaves also contain a bitter yellow sap that is more difficult to see than the clear, water-retaining gel because it's limited to the area just under the green skin. This sap irritates mucous membranes, which further discourages animal depredations. Interestingly enough, the earliest medicinal use of aloe vera involved not the soothing gel but this bitter, irritating yellow sap, which was extracted, dried, and ingested as a laxative.

Aloe vera is exceedingly easy to keep as a houseplant because it can withstand long periods of neglect. All it needs is a sunny window, good ventilation, comfortable household temperatures, and a sporadic supply of water. An aloe vera doesn't often flower indoors, but it reproduces itself readily by developing small, genetically identical plants around its base. These small plants, called suckers, already have roots, so to create a new aloe vera all you have to do is separate a sucker and put it in its own pot. Eventually this sucker will grow big enough to produce more suckers of its own.

Commercial aloe growers take advantage of this suckering habit to keep their outdoor aloe fields productive. They wait until the leaves of the suckers are about 15 inches (37.5 cm) tall and then transplant them to a new field. The old plants can still have two or three of their large outer leaves harvested every eight to ten weeks during the year or so it takes for the leaves of the transplanted suckers to reach harvestable size.

Before the 1930s, aloe vera was considered nothing more than a folk remedy and was grown only as a southern landscape plant or a northern houseplant. But during the 1930s researchers discovered that the moisture-storing gel from aloe vera leaves could be used to treat X-ray and radiation burns. Then, during the 1940s and 50s, a chemical engineer who had used aloe vera to heal his own severe sunburn worked on developing a commercial ointment containing aloe gel. These early research efforts were supplied by aloe grown in Florida, but then aloe began finding its way into a multitude of modern ointments, creams, lotions, and shampoos, and commercial production shifted to the Rio Grande valley of Texas, where today more than fifteen hundred acres are devoted to growing aloe vera.

While most of our popular houseplants are merely of ornamental or botanical interest, aloe vera offers medicinal interest as well. Medical researchers are still arguing about its value, contents, and effects, but enthusiasts like my former housemate are out there demonstrating how it works. Given the number of little tubes, bottles, and jars I used to keep in my medicine cabinet for first aid, it seemed practical to me to nurture one experimental aloe vera plant instead. I can't claim medical proof of miracle cures, but I have found the aloe's cool, moist gel as soothing to my minor injuries as anything else I've tried.

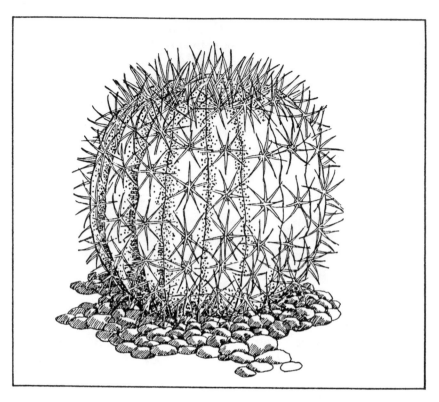

GOLDEN BARREL CACTUS

In trying to understand houseplants and clarify my feelings about them for this book, I visited lots of garden centers, nurseries, and greenhouses. One greenhouse in particular kept me coming back with its unusually large collection of cacti. I found myself dropping by just to feast my eyes on the odd shapes, textures, and growth habits of these most enduring of plants. Despite my resistance to owning houseplants, I managed to convince myself that I wanted a cactus to keep me company in my study.

I surveyed the hundreds of cacti over and over again, trying to figure out a system for telling them apart and then trying to decide which individual was the most representative cactus for someone who wanted only one. Finally, I got so frazzled I had to abandon my attempts at analysis and just pick a cactus that appealed to me. Once I quit worrying about details, the choice was easy. I selected a small round cactus with an impressive array of long yellow spines that made me think of a bright sun. It was a golden barrel cactus, a species that is readily available because it grows easily from seed and is widely produced by commercial growers.

I bought the smallest golden barrel in the greenhouse because, when I looked at some of the bigger specimens, I thought it might be fun to watch my cactus grow. I've since discovered that I've committed myself to several decades of leisurely observation. It takes this species at least ten years to reach 6 inches (15 cm) in diameter—and that's when it gets plenty of sun, water, nutrients, warmth, and space. In a pot in my study, the growth will probably proceed more slowly. In central Mexico, where the golden barrel cactus grows wild, it can live for several hundred years, gradually attaining a diameter of 3 feet (.9 m) and a height of 4 feet (1.2 cm).

My youthful golden barrel is about 2 inches (5.0 cm) in diameter and 2½ inches (6.25 cm) tall, and at present it has thirteen ribs, with a cluster of long, sharp spines raying from little pads along each rib. Several other plants produce spines, but only a cactus produces its spines on these distinct little pads, called areoles. Plant classifiers look to the number and arrangement of spines growing from each areole to determine the species of a cactus. A mature golden barrel, for instance, has nine to fourteen symmetrically arranged spines growing from areoles that are spaced evenly along about thirty ribs.

My golden barrel cactus will never reach its full size indoors because even my brightest window can't offer it as much sun as central Mexico does, and the sun, of course, is what provides the energy for growth. Even in central Mexico, where the sun shines brightly most of the year, the golden barrel cactus grows only during the summer—a cycle it sticks to indoors—because that's the only time it rains. The cactus uses some of the rainwater to produce new tissue, but it must also store enough to support the activities necessary to survive during the rest of the year.

The distinctly ribbed, heavily spined golden barrel cactus is well adapted to the hot, dry climate that characterizes its native habitat. The ribs enable the stem to expand and contract as it absorbs and loses water. Cacti are called stem succulents—as opposed to leaf succulents, such as jade plants and aloe vera—because they store their water in their thick stems. After a heavy rain, the barrel cactus swells until its ribs hardly show, and during a long drought it shrinks until it's all ribs again.

The spines, too, help a cactus deal with the climate. They

are actually modified leaves. Real leaves would be a liability because their broad surfaces would parch in the sun, so the cactus stem, which is green with chlorophyll, has taken over the leaves' job of photosynthesis, and the leaves have become slender, pointed, multipurpose spines. Their most obvious job is to protect the cactus, with its store of water, from being destroyed by thirsty animals. But the spines also trap a thin layer of air around the plant, which insulates it from the full heat of the sun, and they cast a network of shadows that keep the cactus twenty degrees cooler than it would be if it were bare. Finally, they act as drip tips, concentrating the fine droplets of moisture in mist and dew into larger droplets that drip onto the surface of the soil right beneath the cactus, where the shallow roots can absorb the water.

My golden barrel cactus isn't going to demonstrate all its miracles for me, but if I water it properly, offer it nutrients during the summer when it's growing, and provide it with a pot big enough for its roots to expand, I should at least be able to watch it grow. I like the idea of having a plant companion that's slow and steady—that will spend the next ten or twenty years gradually gaining size without flowering or otherwise threatening me with small duplicates of itself. A golden barrel cactus seems to be the ideal houseplant for someone like me, who is fascinated by plants but isn't really comfortable with a lot of indoor greenery, either when it multiplies or, especially, when it dies.

III. OTHER KINGDOMS IN THE KITCHEN

YOGURT

As a naturalist, I find my kitchen the most interesting room in my house. It is alive with activity, not all of which is as visible or consciously chosen as the activities of pets and houseplants. Take the yogurt in my refrigerator, for instance: it is full of microscopic organisms, or microorganisms, that were active until they got refrigerated and would gladly resume their activities if only I would let them operate at warmer temperatures. The "active cultures" advertised on yogurt containers are living bacteria—one-celled microorganisms that are best known as causers of disease. But there are bacteria and there are bacteria. The ones in yogurt perform the useful service of converting perishable milk into sour but longer-lasting and easier-to-digest yogurt.

Bacteria are neither plants nor animals according to accepted definitions, so scientists have placed them in a separate kingdom called the *Monera*. They are exceedingly simple organisms, each consisting of just one cell with a loosely organized central area called a nucleoid because it lacks the binding membrane of a nucleus. The cells come in three basic shapes: spheres (the *cocci* bacteria), rods (the *bacilli*), and spirals (the *spirilli*). Despite their simplicity, however, bacteria have all the needs of more complex life forms. They must find suitable habitats, they must eat, and they must reproduce.

The bacteria that turn milk into yogurt belong to two different species. One is a rod-shaped bacillus called *Lactobacillus bulgaricus*, and the other is a spherical coccus called *Streptococcus thermophilus*. Both are necessary to create yogurt because they interact with each other to produce just the right acidity and flavor in the finished product.

Commercial yogurt production depends on exact proportions of the two bacteria interacting at carefully controlled temperatures. The bacteria are often grown separately and combined in a ratio of 1:1 just before they are added to the milk. An excess of *L. bulgaricus* could lead to too sharp a taste, while an excess of *S. thermophilus* could slow the development of the desired acidity.

Both yogurt bacteria feed by a process called fermentation. Fermentation involves the breakdown of a carbohydrate—often a sweet carbohydrate called a sugar—without the help of oxygen. The bacterial fermentation of the sugar in milk, called *lactose*, produces *lactic acid*, which gives yogurt its sour-milk taste. But as the fermentation proceeds, the two bacteria interact in ways that produce a custardy, edible yogurt rather than plain sour milk. Between them, the two bacteria also convert the original milk protein into a protein that is more easily digested by human beings. And they furthermore produce an enzyme called *lactase*, which helps us digest whatever milk sugar they leave.

These beneficial bacteria do everything they do not to make life easier for us, but to meet their own needs for food. And while they're feeding, of course, they're growing, and when they reach a certain size they begin to reproduce. Being simple organisms, bacteria reproduce quite simply. Each cell merely divides in two by a process called fission. An equal amount of cell material goes with each half, the two new cells feed and grow, and then each of them divides again.

If the yogurt were left at the 110°F (43°C) temperature commercial producers use to make it, it would become increasingly acid as the growing number of bacteria continued to ferment the milk. To keep the yogurt appetizing to human taste, the bacterial activity must be slowed down after eight to ten hours of warm-temperature activity. Both yogurt bacteria stop growing when the temperature drops below 50°F (10°C) and become completely inactive below 40°F (5°C). In a refrigerator, yogurt should remain tasty for at least two weeks, and I've even had some that was still good after a month.

You can make yogurt yourself if you have a bit of fresh commercial yogurt that lists active cultures on its label. The only other ingredient you need is milk—whole, low-fat, skim, or powdered, depending on what you have. To make your yogurt, heat a quart of milk to a full boil, stirring it occasionally to prevent it from foaming over or burning to the bottom of the pan. This preliminary heating eliminates other microorganisms that might compete with or kill the two yogurt bacteria. Cover the hot milk and let it cool until it is just warmer than your own skin. Then stir in a generous tablespoon of commercial yogurt, pour the mixture into a quart jar, cover it loosely, and set it in an oven that has been preheated to 275°F (135°C). The heat will encourage the bacteria to activate, but you want to turn the oven off as soon as you put the mixture in so that they will spend most of their time at a lower temperature. Let the bacteria work on the milk for at least six hours, and then taste your yogurt to see how strong it is. If you want it stronger, let the bacteria work for eight hours, or stronger still, for ten hours.

This method of making yogurt is a modern version of the traditional Middle Eastern method of making new yogurt from old. It's somewhat inexact, but if the commercial yogurt you used as a starter was still alive, you should get an acceptable batch of fresh yogurt—thanks to the cooperative activities of two beneficial bacteria.

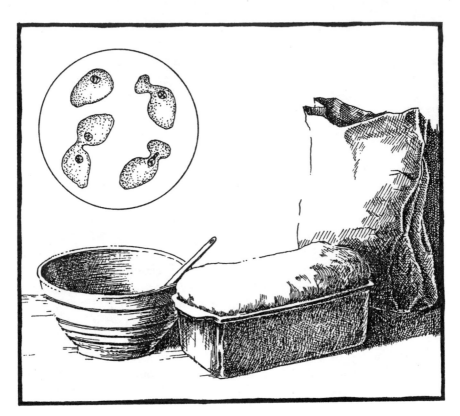

YEAST

Whereas the bacterial fermentation of the sugar in milk produces yogurt, yeast fermentation of the sugars in fruit juices and grains produces wine, beer, and bread. If you have a packet of active dry yeast in your refrigerator, you have billions of individual yeasts to observe. You can't really watch them in action as you would bigger organisms because yeasts, like the yogurt bacteria, are microscopic. But you can certainly see, smell, touch, and taste the effects of their activities on bread dough. And if you are willing to assemble the ingredients and equipment, you can observe yeasts creating beer and wine, too.

But first some scientific background. Baking, brewing, and wine making had been commonplace human activities for thousands of years before scientists took a serious interest in the causes of rising dough and fermenting liquids. It wasn't until 1680 that Anton van Leeuwenhoek examined a droplet of fermenting beer under one of his homemade microscopes and announced the presence of what he called "animalcules." He wrote about them at length, but the scientific community wasn't ready to accept the possibility of life forms they couldn't see. Two centuries later, scientists were still arguing back and forth over whether rising and fermenting were spontaneous, chemical, or biological events. It took Louis Pasteur's studies, one of which again focused on beer (*Études sur la bière*, 1876) to convince people that yeasts were indeed alive and lived in certain predictable ways.

With improved techniques of magnification, yeast research has progressed to the point that scientists can now describe the yeast cell itself, its youthful multiplication of genetically identical cells, and its eventual production of spores. They have distingushed over five hundred different species of yeasts so far, each one with its own requirements and behaviors. Surprisingly, bakers, brewers, and vintners use carefully cultivated strains of just one species in their respective enterprises: *Saccharomyces cerevisiae*. And modern yeast watchers have thoroughly analyzed the processes by which cells of this common species eat and grow, which is what's happening when bread is rising or beer and wine are fermenting.

A yeast, like a bacteria, is a one-celled organism, but it's somewhat more complex in its behavior. It was formerly classified as a plant, but it doesn't have the green chlorophyll that enables typical plants to make their own food, so it, too, has now been classified in a separate kingdom, together with mushrooms and molds constituting the *Fungi.*

In the wild, yeast spores float around in the air, settling onto fruits and into flower nectar, oozing tree sap, and other sugary substances. If the weather is warm enough, the yeasts begin to ferment the plant sugars into alcohol and carbon dioxide. Because fermentation does not require oxygen, the yeast can keep on fermenting even if it is suddenly trapped in a wine cask or beer keg—as our ancestors discovered long before wine making and beer brewing became scientifically exact enterprises.

During this early stage of its life cycle, a yeast of the species we've domesticated for our own uses multiplies by an asexual process called budding. Each cell creates numerous genetically identical cells, which break free and go on to produce more cells themselves. When the budding cells have consumed most of the available sugar, they move into the second phase of their life cycle. They begin to include oxygen in their activities, which enables them to consume the alcohol they've just produced. At this stage, if they're not interrupted by a baker, brewer, or vintner, they create reproductive cells, some of which fuse with each other by a process called conjugation. Conjugation allows new genetic combinations, and spores carrying these new combinations keep yeasts capable of evolving and adapting to the conditions around them.

In baking bread, we allow the yeast cells only the first part of their life cycle. We invite them to feed on the sugars in the moistened flour and sweeteners we set them among, but while they're still fermenting and budding and producing the carbon dioxide that makes bread dough rise, we bake them in a hot oven, which causes a last burst of carbon dioxide production before killing the yeast cells and evaporating any alchohol they produced.

In wine making, we also kill off the yeast cells after they have turned the grape sugar into alcohol. In this case, though, we don't want to evaporate the alcohol, so we don't cook the yeasts at a high temperature. Instead, we cork the cask or bottle and deprive them of the oxygen they need to switch from alcohol production to alcohol consumption. The yeast cells eventually die, imparting a pleasant flavor to the wine.

Beer brewers keep their yeast cells alive for future use, removing them from the brew after they have produced the desired levels of alcohol and carbon dioxide. This brewer's yeast can be used again for brewing or baking or converted into fodder yeast for livestock. It can also be processed into the nutritional yeast you can buy at health food stores.

It might be difficult to imagine the activities of invisible yeasts, but whenever you eat bread or drink wine or beer, you're enjoying the managed byproducts of their lives.

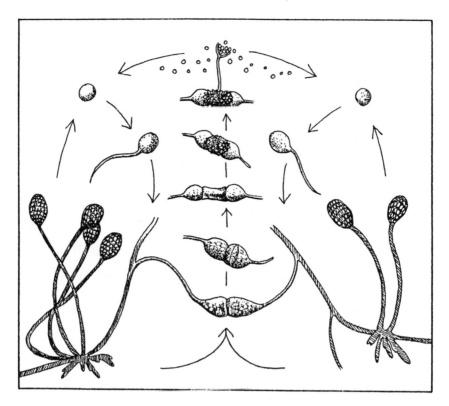

BREAD MOLD

Other fungi, more visible than yeasts during the advanced stages of their life cycles but just as microscopic upon their arrival, often find exactly the conditions they prefer in my kitchen. Some of the most common of these fungi are the ones that cause bread mold. I have inadvertently cultivated some rather impressive specimens of molding bread by leaving heels in their plastic bags and forgetting about them. But I've also been more methodical. One summer I actually watched the progress of a piece of molding bread for my own edification. It's not easy being a microbiologist without a microscope, but observing the very visible work of bread molds offers an intriguing introduction to the field.

"Bread mold" is usually defined as a single fungus, *Rhizopus stolonifer*, which is cottony white when it starts growing and covered with little black specks when it matures. But my piece of bread developed numerous other colors, including blue, green, red, pink, and cheesy yellow-orange. Bread is appealing to at least twelve different species of fungus, five of them quite common, and the different colors reflect the mature stages of the different species.

The microscopic spores that grow into bread molds are everywhere and almost impossible to eliminate. They float in the air and ride around on the surfaces of fruits, vegetables, and cooking implements. Commercial bakeries sometimes add calcium propionate to their bread dough to retard development of molds in their bread, but the spores can outwait the calcium propionate given enough time. Calcium propionate lasts for only four days, after which the mold spores are free to grow.

Some bakeries also use ultraviolet light to kill spores on the cutting equipment and on the cut bread before it is packaged, but once the package is opened, new spores land. Home baked bread, lacking retardants and ultraviolet treatments, is especially vulnerable to mold spores from the minute it cools to a temperature that no longer discourages them.

Each species of bread mold has a slightly different way of growing and producing its spores, but they all have the same relationship to the bread. Like yeasts, they don't have chlorophyll to help them manufacture their own food, so they must obtain their food from other sources. Warm, moist bread happens to be a favorite because it offers a good supply of starches and sugars, but many of the molds that like bread also like fruits, vegetables, and cheeses. They choose bread so often only because it tends to be more readily available to their spores as they float around the kitchen.

When a *Rhizopus stolonifer* spore lands on a piece of home-baked bread, or grocery store bread that's over four days old, it starts growing immediately. If the moisture and temperature are right—as they often are inside a plastic bag in a warm kitchen—the mold grows rapidly, becoming visible within twenty-four hours. First the spore produces a single thread called a *hypha.* This hypha soon starts branching to produce a mass of hyphae called a *mycelium,* which appears as a cottony spot on the surface of the bread. Feet called *rhizoids* descend from the surface network of hyphae into the bread to absorb food for the growing mycelium. Then, after a brief period of growth, the mold is ready to reproduce.

During the first stage of reproduction, the surface hyphae send up little stalks at the points where they earlier sank feet into the bread. Small spore cases develop at the tops of these stalks, and as the spores inside mature, they darken until they are black. Finally, the spore cases break open, each one sending fifty thousand or so sooty little spores into the atmosphere. The single spore that started growing on the piece of bread can, within three or four days, launch millions of new spores.

This first method of reproduction, like the yeast's budding, is asexual, involving the genes of only one parent. The asexual spores are called *conidia* to distinguish them from spores produced by two parents. Sexual reproduction requires that spores from two different parents land close enough for the tips of special exploratory hyphae to meet. If the hyphae are, furthermore, of opposite types—one must be a plus strain and the other a minus—they respond to contact by swelling into special reproductive cells at their tips. These reproductive cells conjugate to form one cell that combines the genes of both parents. This cell develops a thick wall around it and enters a resting period. After one to three months, it reactivates and produces a spore stalk of its own to disperse spores that carry a new combination of their parents' genes.

This particular life cycle accounts for the cottony whites and blacks of *Rhizopus stolonifer*, but the blues, greens, reds, pinks, and yellow-oranges belong to different fungi that have slightly different ways of reproducing. A naturalist with more of an inclination toward microbiology than I have could stay busy for most of the summer identifying and analyzing all the different species of fungi that will develop on a single piece of bread.

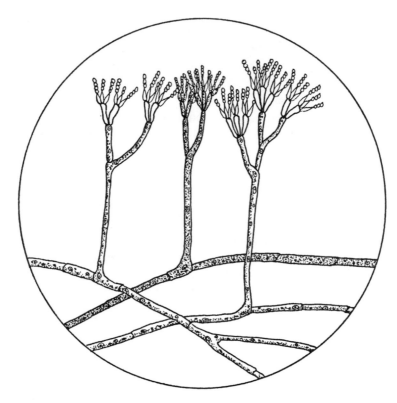

MILDEW

Another group of fungi that hang around my kitchen—and also my bathroom and basement—are those that cause mildew. Mildew is caused by several different species of fungi, some of them closely related to the fungi that cause the blue and green bread molds. Interestingly enough, other close relatives of the mildews produce penicillin.

Finding out about household mildews is somewhat complicated by the fact that certain plant diseases are also referred to as mildews, but they are caused by different types of fungi altogether. Everything that's called a mildew is caused by a fungus, but the downy and powdery mildews that attack plants are parasites: they feed on living plants. Kitchen and other household mildews are, in contrast, *saprophytes*: They feed on nonliving organic matter, such as the oils in paint, the grease on walls, the cotton in dish clothes, the soap film on shower curtains, the conditioning agents added to leather, and even the dust that accumulates on camera and binocular lenses. Many of the fungi that cause these household mildews belong to a special subgroup of fungi called the *Fungi Imperfecti*—"imperfect" because no researcher has ever observed a "perfect," or sexual, stage in their life cycle.

The life cycle of one of these imperfect, saprophytic, mildew-causing fungi begins—insofar as a cycle begins anywhere—when a spore finds a suitable place to germinate. Like bread mold spores, mildew spores are always drifting around in the air, finding their way into houses, and settling indiscriminately onto indoor surfaces. All they need is sufficient moisture to invite them to germinate, enough warmth to encourage them to grow, and a very little bit of food to support their microscopic activities.

A germinating mildew spore produces hyphae that grow and branch in search of food. If these initial hyphae don't encounter some organic matter within about twenty-four hours of germination, the mildew dies before it produces a mycelium. But if the hyphae make contact with something organic, they start feeding, and if they are not interrupted, the mycelium grows to a considerable—but still microscopic—size.

As the mildew feeds and the hyphae continue to grow, this imperfect fungus enters the reproductive stage of its life cycle. Some of the hyphae become specialized stalks and produce conidia at their tips. Whereas most other fungi, like the common bread mold, move beyond this simple, asexual phase of reproduction to a sexual phase during which they mix genes with the genes of other individuals, mildew fungi stop here. Nonetheless, the asexual spores soon achieve sufficient numbers to become visible as spots of mildew, which is when their interests begin to conflict overtly with ours.

Wiping a spot of mildew away with a damp sponge merely spreads the spores, often leaving the mycelium in place to make more spores. The best way to combat household mildews is to deprive them of the conditions they need to get started. Keep your windows open to circulate fresh air, use fans, and perhaps even install an air conditioner or dehumidifier to reduce moisture, which is what invites mildew spores to germinate. Be especially alert during warm weather because warmth combined with moisture encourages not only germination but rapid growth. Take mildew-prone goods outdoors occasionally to discourage the mildews, many of which thrive in darkness but die when exposed to sunlight. Finally, keep threatened goods clean, and store them in dry, cool, well-ventilated places.

If you discover mildews growing around your house, you don't need a battery of chemicals to eliminate them. A solution of one cup of bleach in one gallon of water will kill the mildews that grow on walls and floors and get rid of the musty odors that are a byproduct of their life processes. A solution of vinegar and water will achieve the same end by creating an environment that is too acid for the mildew. For mildews on upholstery, mattresses, and leather products, use a solution of half rubbing alcohol and half water, and for mildew stains on washable fabrics, use a combination of lemon juice, salt, and sun.

The fungi that cause mildew can never be eliminated completely—there are too many of them and their spores are too ubiquitous—and, anyway, it would be a serious ecological mistake to try. Outdoors these same fungi perform an essential service. Whereas the mildew spores that germinate in our houses become nuisances and threats to our domestic goods, the mildew spores that germinate outdoors function as decomposers. They break down the chemical bonds of organic debris and release elements back into the soil to be used again by other living organisms.

In my campaigns against household mildews, I am only defending my own interests against their misplaced efforts at decomposition. I guess housecleaning is, at its most basic, the human species' attempt to maintain control over indoor environments. But the best we can do is to create temporary environments where mildew and our other adversaries can't find what they need to prosper.

ROTTEN APPLES

In my experience, apples behave in two completely different ways: Sometimes they rot and sometimes they don't. Shortly after I watched my piece of bread mold, I decided I'd like to watch an apple rot. I cut a fresh apple in half, ate one of the halves, and set the other on the kitchen windowsill so I could observe the process of rotting. Nothing happened. Day after day it looked the same, but then I noticed that it was shrinking. Ever so slowly, and with perfect symmetry, the apple shriveled until its rounded surface looked like deeply weathered skin. It became leathery and as light as cork, eventually shrinking small enough to fit into a 1½-inch (3.75-cm) magnifying box. I later learned that I had merely *dried* my half apple—that is, eliminated the moisture that encourages rot—which is one of the oldest methods of preserving fruit.

But I had an entirely different, unintentional experience

with some other apples. I store most of the fresh apples I pick in the fall in my cellar, but I like to have a few upstairs in my kitchen for ready eating. These I keep in a peck basket in the cool cupboard under my sink. I must have pushed the basket to the back of the cupboard to make way for other things before I finished all the apples in it because, when I looked for the basket several months later, I discovered three sad little remains at the bottom. What had been juicy, round, red Northern Spies—with only a few brown spots when I put them in the basket—had become dry, flat-bottomed blobs, re-duced to the size of golf balls and covered with multicolored spores. I didn't get to observe my forgotten apples in the process of rotting, but I know what happened to them. These blobs offered perfect evi-dence of what the bacteria, yeasts, and fungi that rot apples can do if left to their own devices in a dark, moist environment.

Many different species of bacteria, yeasts, and fungi rot ap-ples. Each of these microorganisms has its own preferences and life requirements, and not all of them even get a chance to start growing these days because of the sophisticated storage procedures designed to thwart them. But the microorganisms that do find an apple and begin growing into its tissues find themselves in a complex competi-tion with each other and larger animals over who gets the fruit.

To appreciate the competition and the rot-causing microor-ganisms' strategy for winning, it is first necessary to understand the apple. The apple tree's strategy is to produce a fruit that is tempting to animals who move from place to place. When the apple seeds are mature and ready to travel, the fruit suddenly increases its rate of respiration and produces a substance called ethylene. Ethylene moti-vates physiological changes that transform a hard, green, relatively odorless and acid-tasting fruit into a plump, red, aromatic, and sweet-tasting delicacy. The ripe apple is now ready to be eaten in order to have its seeds deposited some distance from the parent tree.

But the changes that make a ripe apple tempting to a roving animal also make it tempting to many microorganisms. Spores float around the apple orchard and settle onto the apples all summer. Some germinate on immature apples, but most must wait until the apples ripen before they can proceed. Some of these more patient spores germinate when they land but remain inactive until the apple ripens. Others must wait until the apple is ripe before they can even germi-nate. If an apple is wounded, it becomes a prime candidate for certain species of microorganisms, but even if the apple is perfectly intact, other microorganisms can enter through the microscopic breathing spaces in its skin, called lenticels.

Once a rot-causing microorganism starts growing, its strategy, according to ecologist Daniel Janzen, who has studied the interactions of these microorganisms and other life forms, is to make the apple undesirable to animals who would customarily eat it. If the microorganism succeeds in making the fruit look, smell, or taste bad enough, it has won. All it needs to do thereafter is battle with other microorganisms and insects over an apple that larger animals reject as rotten. A microorganism that can compete effectively enough to feed and reproduce itself has led a successful life.

But to return to my apples, even having had their fruit consumed by numerous rot-causing microorganisms, some of which produced extremely colorful spores, they were still doing OK. I cut each of the dry, flat-bottomed blobs down the middle to see what they looked like inside. Whatever colors festooned the mummified remains—representing successful microorganisms—the seeds at the center of each apple remained intact. The rot-causing microorganisms obviously wanted to eliminate their competitors, but they would have defeated themselves if they had also eliminated the possibility of future apples.

IV.   HOUSEHOLD
ECOLOGY

FRUIT FLIES

One animal that enters right into the competition over rotting fruit is the fruit fly. But this speck-size insect is not interested in the fruit—it's interested in the rot-causing microorganisms. Fruit fly larvae relish the yeasts that ferment and decompose overripe fruit. If you leave apples—or pears, peaches, nectarines, bananas and other fruits—out of the refrigerator and they begin to soften and turn brown, your kitchen becomes a fruit fly paradise.

   The fruit fly that inhabits human kitchens should not be confused with the Mediterranean fruit fly, which has caused serious problems for fruit growers in Florida and California. This agricultural pest belongs to a different family of fruit flies, the members of which are known for their attacks on fresh fruits. The female of the Mediterranean fruit fly rams her stiff ovipositor into the rinds of healthy, growing oranges and melons to lay her eggs, and the larvae eat their way into the pulp, making the fruit unmarketable.

The female of the household fruit fly, in contrast, uses her telescopic ovipositor to push her eggs into soft, rotting fruit. Her eggs, in fact, have special respiratory tubes that enable them to incubate in the moist brown spots the female chooses for them. When the eggs hatch two days later, the whitish larvae feed on the yeasts that are causing the surrounding brown spot. Then, after three to five days of eating yeasts, the full-grown larvae creep to the surface of the fruit to pupate in a drier environment. Four or five days later, the mature adults emerge, winged and ready to fertilize or lay another two hundred or so eggs. During the warm months of summer, it takes only ten days to two weeks for a fruit fly to grow from egg to adult, so a bowl of rotting fruit—or a garbage pail or compost heap—can nurture numerous generations.

This prolific fruit fly of the kitchen also thrives in scientific laboratories under more controlled conditions. There it lives in glass jars, provisioned with wet oatmeal, brewer's yeast, rotting bananas, and other laboratory concoctions. Even in a glass jar, its reproductive rate is rapid and prodigious, making this species an excellent subject for studies of genetics. Conveniently, the cells in the larva's salivary glands contain giant chromosomes—only four pairs, as opposed to the twenty-three pairs in human beings. With a microscope, a scientist can readily determine the locations of different genes along the length of the ribbonlike chromosomes.

In some of the earliest studies of fruit fly genetics, which began in 1909, a researcher named Thomas Hunt Morgan noticed a white-eyed fruit fly among numerous red-eyed siblings. He mated this aberrant white-eyed male with a typical red-eyed female, and then mated brothers and sisters of the next generation until he produced more white-eyed males. He then studied the chromosomes and figured out exactly which gene caused the white eyes. After years of studying fruit flies, Morgan succeeded in mapping all their genes and explaining which genes were responsible for which traits. His work, for which he received the Nobel Prize in 1933, produced information that has been useful for studies of plant, animal, and human genetics.

Other scientists have used fruit flies to study evolution. By isolating one population from another and keeping them separate over many generations, researchers have produced species that can no longer interbreed, which is what happens in nature when populations become geographically isolated for a long enough period of time. Because fruit flies living in a warm laboratory can produce as many as twenty generations a year, they exhibit evolutionary trends and developments more efficiently than do other animals.

It may be difficult to imagine creatures the size of fruit flies being able to tell each other apart, but they must if the different species are to survive. The males of each species perform a love song that the females can distinguish and respond to. The "song" consists of wing vibrations emitted at a certain pitch and repeated a certain number of times per second. The kitchen fruit fly, for instance, vibrates his wings at a pitch a few notes below middle C on the piano, and he repeats this sound twenty times a second.

A male fruit fly will court any female who attracts his attention, but the female rejects males who sing the wrong songs. If a male sings the right song, however, she stays still and allows him to mate with her several times. This mating system obviously works because there are over a thousand different species of fruit flies, of which the common fruit fly of the kitchen and laboratory is just one.

Some insects are among our greatest adversaries in matters of food and health, but others, like the fruit fly, are harmless and happen to be among our greatest allies in learning about life. I sometimes think that insects—in their multitudes and variety, in their harassments and contributions—may have more to teach us than we will ever learn.

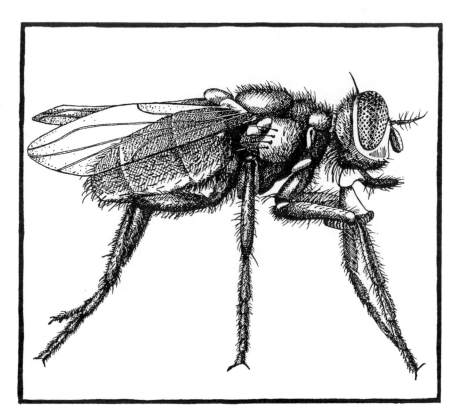

HOUSEFLIES

Houseflies are bigger and easier to observe in action than fruit flies—
and cause more trouble if you let them. My experiences with houseflies extend all the way back to my childhood, when my grandmother
used to offer me a penny for each one I could eliminate. I can still
remember chasing those elusive gray-black creatures all over her big
farm kitchen, waiting for one of them to land on something that
wouldn't break when I swatted it. Because I had been told that houseflies carried germs, I undertook my attacks upon them with missionary
zeal, but, at the same time, I couldn't help being fascinated by them.

They always seemed to know when I was about to swat
them, for instance. I would sneak up on one which had settled on
the kitchen table and carefully raise my fly swatter. Then, just when
I swatted, the housefly would take off. A housefly's antennae are
extremely sensitive to movements of air, so it doesn't have much

difficulty noticing the sudden whoosh that accompanies the descent of a fly swatter.

I was also amazed by a housefly's ability to land on the ceiling and walk upside down. It's impossible to watch what actually happens when a fly executes this maneuver, but English photographer Stephen Dalton managed to capture the whole process on film. The fly senses when it's approaching the ceiling, reaches up to touch it with its front legs, flips the back part of its body forward in a half somersault, and lands upside down on the ceiling. The fly can walk on the ceiling—the same way it can walk up walls and windowpanes—because it has sticky little hairs on its feet.

These sticky hairs contribute to the misery houseflies can cause human beings. Unfortunately, in addition to helping houseflies defy gravity, the hairs sometimes accumulate pathogens—the types of microorganisms that cause diseases. Houseflies live in two different worlds—the indoor world of sugar bowls, plates of cookies left on the kitchen counter, and slices of bread waiting for someone to make sandwiches, and the outdoor world of manure, outhouses, compost heaps, and other decomposing matter. They lay their eggs and do a certain amount of feeding outdoors and then fly into the kitchen through an open door or window to crawl over whatever other foods they can find indoors.

And their sticky, hairy feet aren't the only problem. Their way of feeding and their digestive systems can spread pathogens, too. A housefly's mouthparts are designed to lap up liquid foods. Outdoors, they lap up the nutrient-rich moisture associated with decomposing matter, but indoors they encounter such dry foods as sugar, cookies, and bread. In order to eat these dry substances, they must regurgitate a drop of liquid, which contains whatever pathogens they've recently ingested, and then lap some of this liquefied food into their mouths.

To compound their offenses, houseflies who have eaten too much often regurgitate some excess liquid without bothering to lap it back up. If you leave a book or magazine open on the kitchen table, a housefly sometimes spots the pages with these little yellowish regurgitations. And, finally, the pepper-sized black "fly specks" that houseflies leave behind contain whatever pathogens survived the trip through the insect's digestive system. Some of the diseases that houseflies have been known to transmit—which diminished considerably after horses and outhouses were replaced by cars and indoor plumbing—are typhoid, dysentery, and cholera.

As with all insect pests, the best way to control houseflies is to understand their habits and life cycle and attack them from all directions. Houseflies lay their eggs outdoors in warm, moist decaying matter because that—and the microorganisms therein—is what their young like to eat. Within a day or so, the eggs hatch into legless white larvae called maggots that eat and grow enough to molt three times during the next four to five days. For the next stage of their life cycle, their final larval skin turns brown and toughens into a protective shell, inside which the insect pupates for another four to five days before emerging as a winged adult. The only way to discourage houseflies during these early stages of their lives is to cover or eliminate the decomposing matter they like to live in.

When the adults emerge as full-grown houseflies, ready to buzz into your house in search of a random meal, you can thwart them with screens. Any that do get in, you can swat with your trusty fly swatter. If you manage to kill a young female, you may be eliminating as many as a thousand offspring, and swatting older females and any males also reduces the local potential for reproduction.

Covering your food will discourage houseflies from spending time in your kitchen, and hanging sticky flypaper will offer them a lethal alternative. Yet another control that I have discovered as a result of my own haphazard housekeeping is my resident spiders. Wherever I find a spider's web suspended in some quiet corner, I always find the remains of houseflies scattered beneath it.

Whether you use spiders, flypaper, or fly-swatting grandchildren to help you combat your houseflies, it's important to keep these particular insects away from your food. I wish that their habits weren't threatening to our health, because they are probably the most commonly seen, readily recognized, and easily observed insects in North America. But since it's them or us as far as our kitchens are concerned, it's appropriate for us to use our intelligence against their instinctive behavior. When I was a child, I thought my grandmother's pennies were good economy, but now that I am in charge of my own household, I can see that her pennies were also teaching me the beginnings of a good ecology.

ANTS

Ants, like houseflies, travel back and forth between the indoors and outdoors, but their habits and food preferences do not involve the transport of pathogens. The ants I see most often in my kitchen, commonly known as little black ants, live in my burrows under the flat stones of my front walk. Small craters mark the entrances to their underground homes.

Outdoors these little black ants are scavengers, eating whatever bits of plant and animal matter they can find within foraging distance of their burrow. They scavenge indoors, too, eating whatever bits of human food I leave lying around. But they do have preferences, I've learned. One spring I celebrated the first ant parade that found its way into my kitchen by allowing it to do whatever it wanted to. A steady line of ants filed from a crack just below a windowsill to the corner of my sink, where I keep a small drainer of vegetable wastes for my compost heap. Another line was soon headed just as steadily in the other direction.

The ants moved rapidly, their elbowed antennae tapping the surface just in front of them. They were following a scent trail—a chemical path laid down by the first ants who found the food—leading from their burrow to the food in my drainer and back again. Each ant perceived the scent with its antennae. The ants also used their antennae to explore the food once they got to the drainer and to explore each other if they happened to meet in their travels.

Because my drainer contained too many odds and ends for me to see exactly what it was the ants were after, I set up a feeding station to determine their food preferences. I turned a dinner plate upside down, and on its rounded, easily accessible surface I dabbed a few items: a little peanut butter, some honey, some cottage cheese, and plain water. As the day went by I added a piece of apple, some raw egg white, milk, and red wine.

Every hour or so I checked my feeding station to see what the ants were doing. I soon learned that they were interested in everything except the wine, but their real favorite was the honey. There were always four or five ants crowded around the honey drop, while just one or two were exploring the other substances.

As much as I enjoyed watching the ants making their food choices, I concluded that such encouragement wasn't good for our relationship. I'm willing to tolerate occasional ants, and even a parade if I'm in the right frame of mind, but I can't allow my kitchen to become the major foraging ground for all the ants who live near my house. So I cleared up my feeding station and began my annual effort to eliminate or protect the foods that invite ants to forage indoors.

The ants who paraded into my kitchen were workers—infertile females whose job it is to provide food for the rest of their colony. Ants, like bees and wasps, are social insects, dependent on the cooperative efforts of their entire colony to survive. The workers which forage for food are the most observable members of an ant colony, but back in the burrow are two other castes (or social classes), young in various stages of development, and numerous other workers.

The castes that remain at home include queens and males. These reproductive castes include the single egg-laying queen, who founded the colony and never leaves the burrow; the new young queens, who are raised by the colony and leave when they are ready to mate; and the males, who hatch from unfertilized eggs and will depart when the young queens do. The reproductive castes do not participate in the work of the colony, but, like the young, they need to be fed.

The workers who stay at home do not look any different from the workers who forage, and they labor just as constantly. They feed and groom the reproductives and transport their developing siblings from one part of the burrow to another as temperatures and humidities change. If you've ever disturbed an ant colony, you've seen hundreds of these busy creatures scurrying around with white objects in their mouths. The white objects are not "ant eggs" but helpless larvae and pupae, who would perish if the workers' response to disruption were to save themselves.

An ant society functions cooperatively, not because each ant makes moral choices about the welfare of other ants, but because chemicals moving throughout the colony motivate cooperative behavior. Each worker who visited my kitchen arrives back at the colony bearing a bit of food in its crop—a storage compartment that is separate from its digestive system. When it encounters a member of its colony who stayed at home, the two will stroke each other with their antennae to be sure that they bear the scent of the same colony, and then they will regurgitate a drop of liquid from their crops and share it with each other—a process called *trophallaxis*.

While the returning forager offers food from outside the colony, the worker who stayed at home offers chemicals it has licked from the bodies of the larvae and egg-laying queen. These chemicals motivate the various ants to keep performing their various tasks, all for the welfare of the colony.

All in all an ant's life is not a bad life, if you can accept the premise that the colony is more important than any one individual. Ants have no trouble with this premise because it is impossible for an individual ant to function without the chemical signals that motivate its behavior. As I watch the little black ants parade into my kitchen each spring to find food for their reactivating colony, I sometimes envy them the simplicity of their chemically motivated lives.

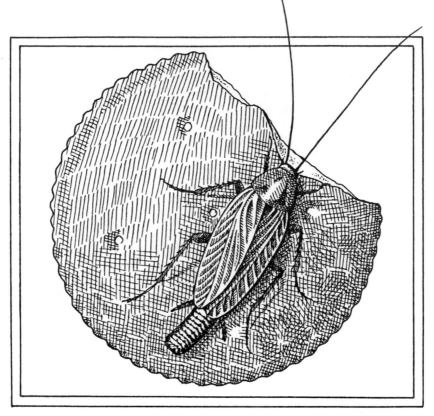

COCKROACHES

I have not seen a cockroach since I moved to Vermont, but I saw enough of them during my years in Washington, D.C., to last me a lifetime. They have a way of making an indelible impression. I can still remember my revulsion and all my frustrated attempts to eliminate these insects from my city apartment, but had I been a naturalist back then, I would have assumed an entirely different attitude.

Unlike houseflies, who travel back and forth between pathogen-rich matter and human food, cockroaches tend to stay in one place. They are therefore less likely to transmit diseases than houseflies are. These flat, brown insects are actually much cleaner than most people think. If you could observe cockroaches without scaring them into their dark hiding places, you would see that each individual spends much of its time grooming itself. First, it rubs its long legs over various parts of its body, combing and scraping away dust and debris. Then it runs its legs and whiplike antennae through its jaws. Because the most common indoor cockroach, the German cockroach, tends to live in a human kitchen, eating human food—with a special preference for breadcrumbs and beer—it is basically as clean as its human landlords.

German cockroaches are essentially scavengers, who have been drawn to indoor environments by the convenient combination of warmth, moisture, and food. In the tropics, where they originated, they hid under moist vegetation during the day, for which they have substituted the moist environments under kitchen sinks and dishwashers. And at night they crept out to feed on whatever they could find in the way of plant and animal detritus, for which they have substituted human food.

Despite its European name, the German cockroach is a native of Africa. But the species has managed to find its way around the world aboard ships. It was first noticed in the United States during the late 1800s, when water from the Croton Reservoir was piped into New York City. The piping gave German cockroaches runways by which to travel, and they followed the new water supply right into apartment buildings. They then followed plumbing networks from apartment to apartment, readily distributing themselves throughout entire buildings. Because their appearance coincided with the construction of the Croton Aqueduct, New Yorkers nicknamed them "Croton bugs."

Despite this particular dispersal, German cockroaches are not in general travelers. They prefer to stay in one location, if they can find enough food to support their expanding populations. Males, females, and young all emit a scent, called an aggregation pheromone, which encourages groups to gather in the same safe hiding place day after day. These aggregations disperse only when they run out of food or when a chemical repellent applied in one apartment drives them next door.

In addition to these aggregation pheromones, which travel a considerable distance, German cockroach females also emit a sexual pheromone, which doesn't travel at all. Given the close quarters this species prefers, a male is just as likely to touch a female as to smell her from a distance. And once he has touched her with his antennae, her scent inspires him to mate.

The mating of German cockroaches requires a careful coordination of movements and actions, all of which follow from the initial antennal contact. After the male has determined that the female is receptive, he turns his back to her and raises his wings. She then climbs onto his back to get at a secretion he produces from glands under his wings. The female must mount the male in this fashion for the male to be able to grasp her genitalia with his. Once he's made the connection, he twists out from under her and the two stand end to end until the mating is complete.

When the mating is accomplished, the male goes off in search of another female, and the female begins to develop an egg case to hold her thirty or so eggs. She carries this egg case around as what looks like a stubby tail for three to four weeks, until a day or so before the young cockroaches are ready to hatch. Then she deposits it in a safe place and leaves the emerging young to fend for themselves.

Cockroach young hatch looking like miniature versions of their parents, only they don't have wings. Young German cockroaches molt five to seven times over the next few months, finally becoming winged adults ready to mate. Under optimal conditions this prolific species can produce three generations a year, which, with each female producing multiple egg cases, could mean as many as 400,000 individuals.

Threatened by such statistics, insect researchers have been working on new methods for eradicating German cockroaches, or at least controlling local populations. Chemical insecticides fail because a few resistant individuals survive every assault, and within a few generations they have produced whole populations that are resistant. The most promising approaches to date involve insect growth regulators, chemicals that interfere with normal development. But these hormonelike substances must be sprayed onto individual cockroaches or added to their food to be effective. And each invasion must be treated anew.

Because I was not a naturalist back in my city days, I missed my chance to observe cockroaches with unprejudiced eyes. But I don't think there's any danger that cockroaches will be eliminated anytime soon. I sometimes think it might be worth a trip back to Washington, D.C., just to take a close look at some of the resourceful descendants of my old cockroaches.

HOUSE MICE

My dealings with mice began in Maryland and continued in Vermont, and have involved two very different kinds of mice. The first was the ubiquitous house mouse, a group of whom had taken up residence in the kitchen range that came with the house I rented in Maryland. I didn't know they were there when I moved in, but the first time I tried to use the oven, I discovered them. I don't think I baked any mice, but the lining of the range was so thoroughly saturated with their urine and droppings that the whole house reeked of their presence. For several weeks I baked nothing and fretted about what to do. Then, in a determined flurry of activity, I took the range apart, pulled out the old lining, installed a new one, and set half a dozen snap traps.

Back in those days I had no particular interest in animals, no curiosity about why house mice might choose to live in a kitchen range, and no larger questions about how this species had become so closely associated with my own. It took me several years, but I finally got around to wondering about house mice—after I had learned about wild native deer mice in Vermont. Knowing what I know now, I still would have snap-trapped my Maryland house mice, but I might have taken some time first to observe them and their activities.

House mice have been associated with human beings since the early days of agriculture. The ancestral species inhabited the dry grasslands of central Asia, where they subsisted on seeds and insects. When human beings moved into this area and began cultivating and storing great quantities of seed crops, house mice seized the opportunities offered by this beneficent species. They built their nests in the predator-free environments of granaries, barns, sheds, and houses. And they learned to eat seeds that had been processed into bread, as well as whatever other foods human beings left lying around. A few house mice traveled with human beings wherever they went and established new populations near new settlements.

House mice have been able to keep pace with their human companions because they are physically adaptable enough to survive in most of the environments human beings can survive in, and they are behaviorally flexible enough to mate and reproduce wherever they find themselves. The only places house mice don't do well are where they must compete with established populations of wild native rodents and where they are confined with too little food to support their numbers. Given an absence of competition, ample food, and opportunities to disperse as necessary, house mice set up social systems designed to populate and colonize whatever space is available to them.

In close quarters—in a kitchen or a laboratory, for instance—adult males establish territories, driving out other males but inviting up to ten females to stay. Each male patrols his territory regularly and marks his boundaries with strong-smelling urine. The females, which become responsive to the male when they smell this urine, bear their young about three weeks after mating, each female producing four to seven offspring that need care for about three more weeks. During this period, several females might combine their young in a communal nest and take turns nursing them. When the young mature, many of them, especially the males, are driven out of the parental territory and become potential colonizers of the surrounding area.

If these colonizers are forced outdoors, they adopt somewhat different social patterns. The males are still aggressive toward one another, but they don't establish and defend territories. Rather, the breeding groups move around as food sources change and new habitats become available. These different ways of living and breeding, settling and wandering, assure that house mice are always available to move into whatever spaces become available to them.

Because house mice prefer many of the same foods we do, with the exception that they also eat insects, and because they often try to settle in our homes, they are direct competitors for our resources. Therefore, if a family group tries to establish itself in your kitchen, you'll need to respond to protect your interests. If I had known a little more about house mouse habits, I would have done a little more than replace the lining of my kitchen range and set half a dozen snap traps. For instance, I would also have put all my food in tin cans or glass jars, and blocked the holes they used to get into and out of the range. Mouseproofing a kitchen is a much more effective way to discourage house mice than leaving the kitchen tempting to this opportunistic species and trying to trap all the new individuals who will eventually find their way in.

I'm not sure I would have developed the fondness for house mice that I have come to feel for my Vermont deer mice. But I could at least have learned something about how animals prosper in association with human beings if I had paid closer attention to those opportunists who had settled into my Maryland kitchen.

DEER MICE

When I moved to Vermont, my life changed, and my experiences with the wild native deer mice I soon encountered are emblematic of the transition. When I first heard the scurrying and scratching of mice in my house, I thought I had house mice again, and I prepared to launch the aggressive counteroffensive I had perfected in the battle over my kitchen range. My primary concern was to keep them out of the new range I had just bought, so I set several traps on the floor around it. But I didn't catch anything. After a few days—or nights, rather—of observation, I discovered that the mice I was now dealing with were not interested in my range. They spent most of their time in walls, cupboards, drawers, closets, and other cooler places.

Then I began finding little caches of prune pits, dry beans, and hazelnuts around my house, something I hadn't noticed with the house mice. Sometimes I found just the empty pits or nutshells on my kitchen counter in the morning, each with a tidy round hole where small rodential teeth had gnawed through the tough outer covering to get at the food inside. Other times I found little stores of uneaten seeds that the mice either hadn't gotten around to eating yet or had forgotten. These mice were wandering far from my kitchen, I discovered, hiding some of their food between blankets I had stacked in the guest room closet three rooms away and some in a dresser full of out-of-season clothes upstairs. I was impressed by the mobility and providence of these new mice.

Because deer mice are wary—as befits a wild animal—and strictly nocturnal, I never saw them at work. I decided to live-trap one to see what it looked like. At that point I was even entertaining the notion of allowing these invisible mice to live indoors with me in accordance with my new live-and-let-live philosophy. I bought a small trap that looked like a cage at the local hardware store and set it in the cupboard under my kitchen sink.

When I caught my first deer mouse, I was amazed at what an attractive animal it was. While a house mouse is a sleazy gray all over, a deer mouse is deer-colored above and white below. Its tail is bicolored like its body and has a few hairs growing on it. A house mouse, in contrast, has a naked, scaly tail that is about the same dark color top and bottom.

When I looked at the deer mouse trapped inside the cage, I was entranced by its huge eyes and big ears—both adaptations to its life in the dark. I perceived a shyness and vulnerability in this wild mouse that was totally absent in the beady-eyed house mice I had battled before.

I decided I would live-trap all the deer mice that chanced into my house and take them back outdoors where they belonged. But I had another lesson to learn about how these animals live. I carried each mouse I caught all the way down to the foot of my hill and let it go in the woods near a river. After several weeks of live-trapping, however, I began to feel overwhelmed by the huge population of deer mice my house seemed to support. I was catching at least one every night, and the numbers were adding up.

It occurred to me that maybe the same mice were coming back, but that meant these small creatures were managing to find their way back to my house from over a mile away. I looked up deer mice in a textbook on mammals and discovered that they have a strong homing instinct. They had been known to find their way home from more than two miles away.

Although I had nothing but admiration for these persistent animals, their nightly raids on my food—including their ability to gnaw right through plastic lids—were a problem, and their droppings in my kitchen drawers and cupboards were a threat to household hygiene. So I set two snap traps for several nights in a row and discovered that my total resident population was only seven mice. After trapping that batch, I wasn't bothered by any additional mice for quite a while.

During the summer, I've since observed, deer mice aren't a problem. They don't seem to be interested in being indoors. But in the fall, when cold weather arrives, they begin to seek out snug, dry places to spend the winter. I would be happier if deer mice would never find my house, because I don't enjoy killing them. But I rationalize that I am just another of the many hazards a wild deer mouse has to contend with during the year or two it might manage to survive. If it had stayed outdoors, foraging for food in the woods and storing it in hollow logs, it would have had owls, foxes, fishers, minks, weasels, otters, coyotes, and bobcats threatening its life. Indoors it has only my traps—and, of course, my cat before she died. When I catch a deer mouse in one of my snap traps, I always return its body to the food chains it was intended to be a part of by emptying the trap outdoors.

Although I have to trap the deer mice to protect what I consider my legitimate interests, I don't feel any of the same disgust toward them that I felt toward the cockroaches I encountered in my city apartment and the house mice I encountered in my Maryland house. Deer mice just don't seem as aggressive and intentional in their intrusions. They've taught me that they themselves are the natives here and I am the intruder. They enter my house not because they are opportunistic camp followers but because my house appeared as a warm dry space in the middle of their world.

POLISTES WASPS

Given the number of organisms who hang around in the kitchen, it's easy to think of food as our greatest enticement. But some of the creatures who invade our houses are less interested in food than shelter. Among these animals I think of as wildlife—because they don't depend on me for much of anything—my favorite is the winter visitor called the Polistes wasp.

This brown-and-yellow wasp is sometimes referred to by the common name "paper wasp," but that name risks confusion with other wasps who build paper nests—such as the yellow jacket and the bald-faced hornet, neither of whom you want in or near your house. These other paper makers are much shorter tempered than the Polistes and much more likely to sting you. All three of these wasps are social insects, who will sting in defense of their colonies, but the Polistes is more primitive in its social development and more docile by nature. It poses little threat even to the most curious of observers. And the individuals who came indoors are even less likely to attack than those you see outdoors because they are either males, who don't have stingers, or fattened and fertilized young females, whose primary business is to conserve energy until they can start a new colony in the spring.

I hardly notice the Polistes who sneak into my house in the fall, looking for a safe crevice where they can enter the insect version of hibernation called diapause. They attract my attention more often during the weeks of spring, when a few of them crawl up through the floorboards and wander around my upstairs rooms. Because these individuals are prospective queens, who, if they succeed in founding new colonies, will eliminate lots of insects and caterpillars in the process of feeding their young, I try to encourage them. I put little dishes of honey-water near where I see them and hope they will take enough sustenance to survive until the wildflowers bloom. Adult Polistes feed on nectar.

Among Polistes, a mated female who survives winter and early spring has not automatically won her queenhood. She heads back to the area where she was raised, and there she encounters those of her sisters who also survived the winter. All of them want to build their nests near the ancestral site, so there's a certain amount of aggressiveness and conflict as these survivors decide who will build her nest where—and in some instances, when a group of sisters decides to build a nest together, who will do the egg laying and who will serve as workers. These sisterhoods avoid chaos by establishing a dominance hierarchy, the most aggressive serving as queen, the others ranked below her in accordance with the number of squabbles they've won.

Whether the queen is working alone or has several subservient sisters working with her, the first task of the season is to build a new nest. Although Polistes like to raise their young near where they themselves were raised, they never use the old nest. Like the other paper wasps, they start fresh every year, seeking out sources of the necessary wood fiber from weed stems, dead trees, old fences, and weathered boards on barns and houses. Each wasp scrapes some fibers free, chews and mixes them with her saliva, and creates a mouthful of gray papery building material.

Polistes often build their nests on or near the houses they winter in, which means those individuals you notice indoors may be perpetuators of a local dynasty. Because this species likes eaves, porch ceilings, and other visible spots, you can often observe their progress from a window or doorway. As soon as one cell of the new nest is complete, the queen lays an egg it it. She lays another one in each additional cell until the colony begins to function like a going enterprise.

During the early stages of nest building and egg laying, there is considerable conflict, changing of plans, and shifting around. Some of the solitary females lose interest in what they're doing and abandon their nests. They might join a sister or a group of sisters, which requires a new round of squabbling. A dominant queen might disappear, allowing the top-ranking sister to start laying eggs, and sometimes she disappears, too, leaving the nest to the third in command. It takes until early summer for the colonies to settle into the social patterns that will sustain them throughout the rest of the season.

Meanwhile, the eggs begin to hatch, and the queen, or the queen and her sisters, must feed the larvae. They hunt insects—especially caterpillars—chew them into a mash, and offer the larvae this protein-rich food. The larvae, in return, offer drops of sweet saliva, which is even more nutritious than the flower nectar the adults have been feeding on. After two or three weeks the larvae spin cocoons and pupate, emerging as adults in another two or three weeks.

The early and midsummer eggs produce plain workers, the only Polistes females who will never aspire to be queens. But by August, the colony begins producing bigger, better-fed females—and males—who will fly away from the colony, gather in large aggregations, and mate. The old colony begins to disintegrate about then, and some of the senior members die of old age. The younger workers and the males, who have served their purpose after they've mated, die with the coming of cold weather. Only the fertilized females—next year's candidates for queenhood—might hide themselves inside your house, asking only a safe place to spend the winter before rising to the rigors of spring.

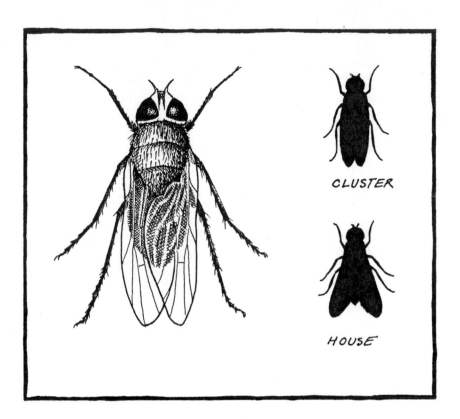

CLUSTER

HOUSE

CLUSTER FLIES

Another insect that likes to spend the winter in human houses is more bothersome than the Polistes wasp. This insect looks so much like a housefly that many people think they've been invaded by bigger, fatter houseflies that have gotten lazy at the end of summer. But these fall invaders, who are called cluster flies for their tendency to cluster together in the hiding places they find—and at windows and around lights when it's warm—are distinctly different from houseflies. If you observe an individual cluster fly closely, you will readily notice the differences.

First, a cluster fly is much slower, clumsier, and louder in its buzzing than a housefly. While a housefly is a spritely, agile creature that easily eludes my fly swatter, the cluster fly acts as if it might be drugged. It bounces off walls and ceilings and crawls so slowly or flies away so lethargically that my fly swatter seems like an unfair weapon.

Cluster flies clustering at the light above my desk have allowed themselves to be massacred—after they've driven me to such irrational behavior with their incessant buzzing—without even trying to get away. I've even had one of these maddening insects land on the piece of paper I was writing on and just stand there as if it were too stupified to move.

If the cluster fly you're looking at is standing still so you can't use its buzzing and bungling to tell it from a housefly, you can still distinguish it by looking at the way it holds its wings. Whereas the housefly holds its wings spread out a bit, making its basic shape a triangle, the cluster fly usually overlaps its wings, making its basic shape more like an elongated oval. Furthermore, the cluster fly's body is covered with golden hairs, giving it a bronze or brownish cast as opposed to the housefly's grayish black.

Another way to tell the difference between cluster flies and houseflies is to pay attention to the seasons. During the summer, while houseflies are in and out of your kitchen after food, cluster flies stay outdoors. They begin to appear indoors only when the weather turns cold in the fall, about the time houseflies disappear. Cluster flies activate on warm days during the winter and start flying against windows in earnest during the early weeks of spring, shortly before you have to start worrying about houseflies again.

This seasonal behavior reflects the cluster fly's annual cycle. Cluster flies are programmed to enter diapause in the fall. Temperatures below 50°F (10°C) motivate them to squeeze into dark hiding places, some of which are cracks leading into houses. When the temperature goes higher, however, the cluster fly is motivated to emerge from hiding and seek light. The insect may have crawled into the crack from outdoors, but it often emerges indoors, where it promptly heads for a window or a light bulb. The furnaces and heaters we turn on to keep our houses warmer than 50°F (10°C) keep cluster flies confused, and many of them exhaust themselves before they find an out-of-the-way corner that's cool enough for them to enter diapause and stay there.

When cluster flies arrive indoors, they are not looking for food. They have stored enough fat to last them for several months of inactivity, and even if all their buzzing around makes them hungry, their food preferences make it difficult for them to feed indoors. Adult cluster flies eat nectar; so even if a cluster fly bungles into your kitchen, it is not likely to steal any food. Nor is it likely to be carrying pathogens, as houseflies do, either in or on its body.

The cluster flies who find safe, cool places to spend the winter reactivate during the early weeks of spring. At that time of year, I perceive them not so much as pests as welcome signs that winter is coming to an end. They mate shortly after they reactivate, and about a month later they're ready to lay their eggs. For egg laying, cluster flies want to be back outdoors again. The females look for moist soil and lay their hundred and fifty or so eggs singly or in small batches, spacing them about a foot apart. By the time a female has finished depositing her eggs, she's distributed them over a sizable area, which gives the larvae who will hatch a few days later maximum opportunities to find food without having to compete too fiercely with each other.

Cluster fly larvae feed on earthworms. When a larva hatches, it is a small transparent maggot, which crawls through natural pores in the soil. When its head runs into an earthworm, it stops, probes around with its mouth hooks, finds a likely spot toward the middle of the earthworm's body, and eats its way right in. It never disappears inside the earthworm completely, however. It always leaves its rear end exposed so it can back out if it wants to.

The cluster fly larva molts three times, feeding inside the worm for the first two stages and emerging to feed on what's left of its outside during the third. The growing larva often consumes the whole earthworm before it pupates. After two or three weeks of feeding (longer or shorter depending on the temperature), the larva pupates inside its last larval skin and about two weeks later climbs out of the soil a winged adult. There are generally four generations of cluster flies during the summer, with the fourth being the individuals who find their way indoors, looking for places to spend the winter.

Cluster flies don't do any damage when they're inside your house, except the psychological damage of buzzing and bungling around your lights when you're trying to read, or dying in unsightly numbers on your windowsills. But I don't think these shelter seekers warrant a chemical attack. Perhaps the best approach to cluster fly control is to tighten your house, sealing every possible crack they might crawl in through. Then use your fly swatter and vacuum cleaner. I always feel a little guilty after I've swatted the cluster flies at the light above my desk or sucked a few into the dark and dusty bag of my vacuum cleaner, but I argue with myself that every cluster fly I eliminate makes life easier for my local earthworms.

HOUSE CRICKETS

House crickets can be either winter visitors, spending the winter indoors and the summer outdoors, or they can hang around indoors year round. The cricket who shaped my impression of this species was a summer resident of my grandfather's library. I spent most of one vacation listening to it and trying to figure out where it was, but I never found it. Crickets are ventriloquists who can shift the position of their wings to change the volume and seeming distance of their songs, and these sudden changes are very good at keeping a human listener confused. Even though I never saw that particular cricket, I am convinced it was a house cricket—Milton's and Dickens' "cricket on the hearth"—because all the good omens and positive associations that have become attached to this species fit perfectly with my grandfather's library.

House crickets were originally residents of warm areas in North Africa and southwestern Asia, but when human beings began to build heated homes in more northerly environments, house crickets expanded their range. They were probably carried to most of their new locations amidst the food and goods of human migrants. A taste for the same foods human beings eat—plus the fabrics we wear and use around our houses—had brought the two species into contact with each other in the first place, and once house crickets had gotten themselves transported north, their need for the warmth of indoor living arrangements created an even closer association. House crickets prefer a temperature of 85°F (about 30°C), which explains their frequent presence around hearths.

House crickets have the choice of living indoors or outdoors during the summer, but they must live indoors during the winter if they are to survive. They can't tolerate cold temperatures, they don't enter diapause, and they haven't evolved seasonal cycles, as have northern field crickets, that would enable them to perpetuate their species by laying overwintering eggs before they died. House crickets are designed to remain active year round, and the only individuals that make it through northern winters are those which come indoors, or happen to be indoors, when winter arrives.

House crickets look somewhat like field crickets, which also find their way into houses upon occasion. But the two can be distinguished in several ways. First, the house cricket is a lighter brown—more of a yellowish-brown than the blackish-brown field cricket. Its lighter colored head also has dark crossbands. If all four of the house cricket's wings are intact, the membranous hind wings provide positive identification. They are much longer than the field cricket's, extending from beneath the leathery front wings with their points reaching beyond the end of the cricket's abdomen.

But a house cricket's rear wings sometimes fall off, which had led some researchers to think that house crickets come in three different types: short-winged, long-winged, and one-winged. One researcher whose curiosity compelled him to examine as many house crickets as he could find, including those in major museum collections, finally concluded that house crickets come in only one type: long-winged. They just lose a wing or two for some unknown reason, leaving only a stump hidden under the leathery front wing where the lost wing grew.

House crickets can survive quite easily without one or both of their rear wings because these wings are nonfunctional anyway.

The muscles that go with them degenerate within a few days of the final molt into winged adulthood, leaving adult house crickets with the broad jump as their major means of locomotion. It's their front wings—the wings they sing with—that are essential to the species' survival. Like a male field cricket, a male house cricket rubs his right wing over his left to produce the jingling chirps we associate with crickets. The two species have a similar song, but the crickets, of course, can tell the difference.

The song our ears can hear is the male's calling song, which invites a female house cricket to find him. Once the female has chosen the male whose calling song is most attractive to her and touched him with her antennae, he shifts to a quieter courtship song. This song inspires the female to climb onto the male's back—a position reminiscent of mating cockroaches—and allow him to attach a packet of his sperm, called a spermatophore, to the underside of her body. After this first mating, the male does his best to keep the female entertained by stroking her with his antennae so she won't leave or eat the sperm packet before the sperm has entered her body. He, meanwhile, is forming another spermatophore and is soon ready to mate with her again.

After several matings, the female seeks out suitable places to lay some eggs before she mates again. Outdoors, she would lay them in moist soil, but indoors she must find moist cracks and crevices. In the laboratory, female house crickets have laid as many as twelve to fifteen hundred eggs during the three weeks of their adult lives. At the warm temperatures house crickets prefer, the eggs hatch in ten to fourteen days, the young looking like miniature versions of the adults. Over the next six to seven weeks, they molt seven to ten times, emerging from their final molt as winged adults ready to mate and lay more eggs. House crickets are very sensitive to slight changes in temperature, however, slowing down all stages of their life cycle considerably if the temperature drops only a few degrees.

To the best of my knowledge, most of the crickets I hear outdoors in the fall are plain old field crickets, trying their best to mate before they die. But I keep hoping a house cricket will show up indoors someday to sing me through a winter, or a summer—I don't care. Given the opportunity, a house cricket can be destructive to food and fabrics, but I'd be willing to risk its appetites if it would make my farmhouse sound as snug and cheerful as my grandfather's library.

SILVERFISH

Another household scavenger, one who is a great deal less acceptable to most housekeepers than a house cricket, is the silent and
secretive silverfish. A silverfish is a difficult animal to observe because
it responds to light by slithering—like a small fish—into a dark crack.
Its sleek little body is only about half an inch (1.25 cm) long, and, if
you try to catch it, it slips right out of your fingers by shedding its
silvery scales.

Silverfish live in human houses because they find the temperatures, humidities, and foods they prefer there. They spend their
days hiding in moist basements or around pipes that create a moist
microclimate. At night they wander into other parts of the house
looking for food. Like many other household insects, they are attracted to the crumbs and food wastes they find in kitchens, but
silverfish also have literary tastes. They relish old book-bindings and

the glazed paper used in magazines. They also like the gum used to attach labels, the glue used to attach wallpaper, and other pastes and starchy substances. Scavengers that they are, they offer small house-cleaning services in return for whatever damage they do by eating dead insects, molted skins, egg cases, and other protein-rich animal detritus.

Silverfish can cause problems if they find their way into libraries, museums, or other places where books and magazines are stored for long periods. With plenty of food and no interruptions, their populations grow to the point that their combined and perpetual nibblings do real damage. In most homes, however, there is enough turnover and activity to limit the damage silverfish can do.

Silverfish have special interest for me as a naturalist because they are primitive insects. Most evolutionists consider them direct ancestors of the more advanced winged forms that evolved after six-leggedness—one of the defining characteristics of the group of animals classed as insects—had become established. The primitive status of a silverfish's six legs is reflected by vestiges of numerous other legs that are still visible as small appendages along its abdomen. Furthermore, both the mating pattern and the way young silverfish grow into adults indicate primitive approaches to reproduction and development.

While more advanced insects copulate, silverfish mate ex-ternally. When silverfish of opposite sexes encounter one another, they first explore each other with their antennae. The male is more active during this mutual exploration, dancing around the female to determine whether she is ready to mate. If she is, he stands in front of her and swings his tail back and forth, spinning a few strands of silk between two supports just above his body level. Beneath these strands he deposits a droplet of sperm.

The female then slithers under the silken strands, which she detects with the long, touch-sensitive appendages that protrude from her rear end—for which silverfish and their relatives are called bristle-tails. The strands of silk let her know that the sperm droplet is nearby, so she stops and feels around for it with her genital opening. After she picks up the sperm, the interaction between the two silverfish is over, and they have no further interest in each other. They resume their respective wanderings in search of food, the female depositing her hundred or so eggs singly or a few at a time in crevices and other safe, moist places over the next several weeks. When she molts, what-ever sperm is left disappears with her old skin, and she's ready to mate again.

When the eggs hatch three to seven weeks later—the incubation time depending on the temperature—the young are at first fat, naked, and milky white. But after a few days they molt and begin to look more like their parents. After the third molt, they are covered with the silvery scales that identify them as silverfish.

For the next year or two, young silverfish don't change much. They merely grow, reaching sexual maturity after about twelve molts. Even after they mature, silverfish continue to molt three to five times a year until they die, which may be at the advanced age of seven years. This gradual and continuing growth, without a metamorphosis into a winged, nonmolting adult, represents a primitive pattern of development.

I don't see many silverfish around my house, but I once caught a glimpse of one in a dark cupboard, so I know they're in residence. While I was writing this book, I tried to trap a silverfish for observation, but I didn't have much luck. Back in the 1940s a researcher discovered that the easiest way to trap silverfish for his studies was to use jars baited with flour. He wrapped adhesive tape around the outside of each jar so the silverfish could climb the vertical surface to get to the flour, but he didn't provide them the same courtesy on the inside. In California, where he was working, hundreds of silverfish climbed into his jars and found themselves trapped by the slippery glass, but my Vermont silverfish were less cooperative. During the six weeks I left my traps set in my study, kitchen, and basement, I didn't catch one.

Because whatever silverfish I have don't seem to be damaging anything I value in their scavengings, I don't worry about them. In fact, I'm glad they're around so that I have representatives of a primitive stage of insect evolution living right in my own home.

SOW BUGS

Yet another indoor scavenger taught me to look more closely at the various insectlike creatures I encounter around my house. Not all of them are insects. An insect, whether it's winged, like a house cricket, or wingless, like a primitive silverfish, has six legs, while some of the other common household arthropods—joint-legged animals with a hard outer body covering instead of an internal skeleton—have more than six. This half-inch (1.25-cm) creature of mine had fourteen.

When I first noticed it crawling around on the floor of my downstairs bathroom, I thought it was a beetle of some sort, but something seemed not quite right. I put the mystery creature into a magnifying box to examine it more closely, and when I counted fourteen legs—seven identical appendages on either side—I knew it couldn't be a beetle because beetles, being insects, have only six legs. But I had no idea what else it might be. I figured I was up against a new type of animal that would force me to learn new Latin names and new anatomical details designed to give an aspiring naturalist a headache.

But I was wrong. When the Extension Agent who helps me identify my mystery creatures told me what I had, I discovered that I knew the type well—I had just never encountered it on dry land. The fourteen-legged sow bug is a crustacean, a member of the same group of arthropods that crayfish, lobsters, crabs, and shrimps belong to. Most crustaceans still live in the watery environments where they evolved, but the sow bug, and its close relative the pill bug, which rolls up like an armadillo when it's threatened, have ventured onto dry land.

Sow bugs don't look the least bit like crayfish, lobsters, or any of the other familiar crustaceans. They look more like miniature, flattened, and elongated turtles. But under magnification, the sow bug's attractively decorated "shell" turned out to be a series of neatly jointed body segments. And the elbowed antennae and all the busy little legs were most unturtlelike.

Sow bugs have been quite successful in their move onto dry land. They have adapted so completely to terrestrial life, in fact, that they drown if submerged in water. They still need moist habitats, however, and seek out the damp, dark spaces under rotting logs, under rocks, and in basements, bathroom, and kitchens. The moisture prevents desiccation and also wets their gills, a functioning remnant of their aquatic past.

Sow bugs, like house crickets and silverfish, are scavengers, but they don't have the taste for fabrics, glue, and paper that makes these other two potential pests. They subsist primarily on plant matter. In the laboratory they have been fed carrots, potato peels, lettuce, and Purina rabbit chow, but in the wild they make do with humbler fare—mostly dead leaves and rotting wood. They eat fairly constantly, processing this decaying matter into yet smaller bits of animal waste, with which they enrich the soil. They can become pests in greenhouses because they nibble at the roots and stems of seedlings, but in the woods—or in a basement, bathroom, or kitchen—they don't do much harm.

To protect themselves from cold winter weather, sow bugs become inactive, but when the temperature rises to above 40 or 50°F (5 or 10°C), they start moving around again, searching for food, mates, and other necessities of life. Female sow bugs begin their annual reproductive efforts in March or April. Once the female has mated, she can store enough sperm to last her through the two or three broods she might produce during the long, warm days of summer. Each brood consists of thirty to forty young, who are incubated for about thirty-four days in their mother's brood pouch.

This unusual structure is on the underside of the female's body, between her second and fifth pairs of legs. It is filled with a watery fluid that bathes the eggs and young while they are developing. By the time the young are ready to emerge, they are miniature versions of the adults, and they walk out of the brood pouch on their own. The female then molts, produces a new brood pouch, and deposits a new batch of eggs in it.

Meanwhile, the first young molt within twenty-four hours of leaving their mother and begin a lifelong process of eating, growing, and molting. Males and nonbreeding females molt about every twenty-eight days, while breeding females molt less often to allow time for the young to incubate. Unlike an insect, who molts all at once by splitting its hard outer covering down the back and crawling out, the sow bug molts in two phases. First it sheds the rear part of its covering, and then, about twenty-four hours later, the front part. While its new rear end is hardening, it still has its front legs to get around on, and by the time its front end goes, it has its rear legs back. The sow bug eats both the rear and front halves of the covering it sheds, recycling the calcium into its new covering.

Researchers aren't sure how long sow bugs manage to stay alive in the wild, but these unusual little crustaceans are designed to last longer than one brief season. Most of them probably live two or three years. I'm not sure of the age, sex, or particular mission of the sow bug I found crawling around in my bathroom, but after identifying it and reading up on its life habits, I released it to pursue its harmless activities—congratulating myself on the discovery that I have a crustacean among my household arthropods.

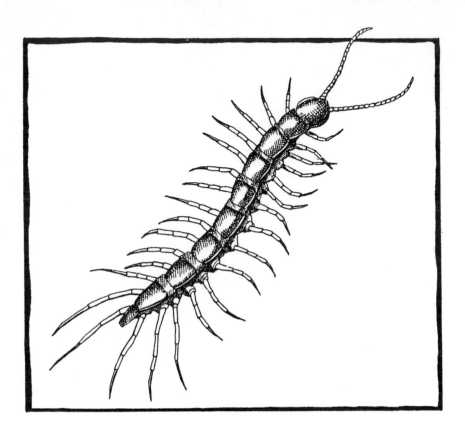

HOUSE CENTIPEDES

Sometimes, against my better judgement, I get to poking around in the dark corners of my basement. I've long since learned that some strange-looking creatures live down there, but curiosity still goads me. Once I rousted out a house centipede, which had me worried until I figured out what it was and what it was doing in my basement.

A centipede is more closely related to insects than sow bugs are, but as its name implies, it also has too many legs. A centipede is characterized by a long, flattened, many-segmented body and lots of legs arranged one pair per body segment. Millipedes, which are often mentioned in the same breath with centipedes but are seldom found in a house, are rounder-bodied, looking more like worms, and have two pairs of legs on each body segment. The common house centipede's flat, grayish body is only about an inch (2.5 cm) in length, but thirty long legs and a pair of long antennae make this small arthropod seem bigger and more threatening than it is.

The house centipede I encountered in my basement worried me because it moved so fast on so many legs, I thought it must be a speedy aggressor, who waited in dark places to jump out and bite human beings. But, when I considered what I had observed, I had to admit that the centipede was actually using its long legs with impressive efficiency to race away from me as fast as it could. House centipedes are indeed aggressive, but not usually toward anything as large as a human being. The few reports of house centipedes biting human beings involve peoples who were trying to pick them up or brush them out of the way. Their more customary victims are flies, cockroaches, and other insects, which the centipedes can eat after they've bitten them.

A house centipede hunts randomly and searches widely for its meals, usually patrolling the house it lives in during the night. It wanders around floors, into crevices, up walls, and across ceilings, using the flat, hairy feet that make up much of the length of each leg to cling to rough surfaces. It uses its long, forward-stretching antennae to sense prey, and the long backward-stretching pair of legs at the other end of its body to sense a rear attack.

When a house centipede comes across a prospective meal, it pounces quite suddenly, using its head and first several pairs of legs to grab the surprised insect. Then it bites the insect with a specialized pair of legs called poison claws. The centipede's poison paralyzes and sometimes kills the struggling insect, after which the centipede manipulates and eats its meal with the help of the poison claws and three pairs of mouthparts.

If you find a house centipede in a sink or bathtub, it has usually fallen from a wall or ceiling during a tussle with an insect. While the long legs and specialized feet are perfectly adapted to moving across wallpaper or even painted surfaces, they are no match for the smooth porcelain of bathroom fixtures. If you try to pick the centipede up by some of its long legs, it will probably shed them—it can regrow them later—to escape you, or if you get a good grip on its body, the centipede may bite you in self-defense. The easiest way to liberate a trapped centipede is to provide it with a towel to climb up and out on.

My house centipedes seem to stay in my basement, or maybe they're just sure-footed enough that they've never fallen into my sinks or tubs during nocturnal forays into other parts of my house. In the basement, where it's dark and moist and they have plenty of places to hide, they live out their predaceous lives and, with luck, produce more house centipedes.

When adult house centipedes of opposite sexes encounter one another in their wanderings, they explore each other with their antennae to determine whether they are both in mating condition. If they are, they may court for as long as an hour, moving in a circle, each touching the other's posterior with its antennae. Only after this preliminary courtship is the male ready to produce a spermatophore. Whereas some species of centipedes spin a small web to hold the spermatophore, the male house centipede merely deposits his on the ground. After he's produced the spermatophore, he pushes the female toward it, encouraging her to pick it up with special appendages she has for this purpose.

When this external mating is complete, the two wander off without further interest in each other. Shortly the female begins to lay her eggs, depositing them singly around the environment she inhabits and leaving them and the young that will hatch from them to their own devices. A young centipede looks somewhat like its parents, but it's smaller and has fewer legs. It hatches with only four pairs, gains another pair after its first molt and two more pairs at each of the next five molts. But even after it has acquired all fifteen pairs of legs, it must molt four more times before it is sexually mature.

Although the centipedes that live in my basement—and perhaps sneak around the rest of my house—can still disconcert me when I see one unexpectedly, I consider them a welcome presence. Along with my more visible and fewer-legged house spiders, they constitute my domestic pest patrol. As long as I have centipedes and spiders in residence, I'll never have to worry about insects taking over my house.

HOUSE SPIDERS

House spiders are the most misunderstood and mistreated of the indoor arthropods. Feared by some and considered a sign of poor housekeeping by others, this quarter-inch (.6-cm) creature is often squashed or vacuumed away with no thought given to the important services it was performing as an indoor predator. Because house spiders, with their webs, are more visible than house centipedes, they are easier to eliminate, but if you consider their greater numbers and the strategic locations of their webs, you should be in less of a hurry to eliminate them. These ubiquitous predators will themselves eliminate many of the insects who fly into your house if you just let them be.

A house spider can be distinguished from your other household arthropods by its eight legs, its two body parts joined by a slender waist called a pedicel, its lack of antennae, and its habit of spinning a web. It can be distinguished from other spiders by its messy, unpatterned web—often referred to as a cobweb—its tan, mottled abdomen, and its preference for houses.

I have house spiders living in every room of my house, but these web-spinners seem to have my habits well analyzed because I've never yet walked into one of their cobwebs. They build in high ceiling corners and other spaces that are out of my way—like behind a door that stays open most of the time, beneath a shelf I don't use very often, between a window and a storm window, and among the pipes and heating ducts in my basement.

One cold winter day, when I was entertaining myself by looking for signs of indoor wildlife, I decided to explore one of these basement cobwebs with a flashlight and a hand lens. I noticed a motionless spider hanging upside down near the edge of the cobweb and assumed that it was dead. But when I leaned closer, my flashlight touched the web, and the spider moved. At first it moved very slowly as if it were just stretching its legs. Then, as I leaned closer still, trying to look at the spider through my hand lens, it shot upward through the tangle of its cobweb as fast as if all those crisscrossing strands were an even surface. Every time I tried to focus my hand lens on the spider, it dashed out of range.

House spiders, as I have learned, do not die or hibernate when winter comes. They don't find as many insects to eat during the winter months, but they manage to find enough to stay active. In their life habits, house spiders are solitary, sedentary predators. Each builds a cobweb and waits for insects to fly into it. When an insect hits the cobweb, its thrashing as it tries to extricate itself from the silk signals the resident spider that a meal has arrived. The small spider approaches the struggling insect, which might be two or three times its size, and throws freshly spun silk over it with the help of combs on its hind legs. With the insect bound in silk, the house spider can easily inject it with paralyzing venom and then eat it at its leisure.

A female house spider spins her cobweb and stays there. A young male spins a cobweb for catching insects while he's growing up, but when he matures, he wanders away to find a mate. When he locates a likely female, he plucks a strand of her cobweb and waits for her response. He has to proceed cautiously because if she's not interested in mating, she might race over and attack him as if he were an insect. If she's receptive, however, she allows him to approach, and if she's still in the mood by the time he gets there, they mate.

The male has already prepared himself for mating before he begins his travels. He deposits a drop of sperm on a special little web he weaves and picks it up with a short leglike structure called a pedipalp. A spider's pair of pedipalps grow between its fangs, or *chelicerae*, and its first pair of walking legs. Both sexes have these pedipalps, which they use to manipulate an insect they're eating, but the tips of the mature male's are specialized for mating. When the male finds a receptive female, he inserts his sperm-filled pedipalp into an opening on the underside of her abdomen, and his mission is accomplished. After the mating, the female might eat the male, or she might just forget about him and let him hang around her web until he dies.

About six to eight weeks later, the female lays her eggs, wrapping groups of them in gray, pea-sized, silk cocoons. She can produce several cocoons after just one mating. The young spiderlings hatch from their eggs one or two weeks later, and from the beginning they look like miniature adults. They grow gradually, undergoing a series of molts as they outgrow their skins. While they're still inside the cocoon, they grow enough to molt once, and some of these growing spiderlings, having finished the bit of yolk provided for them, begin eating their brothers and sisters.

When the survivors emerge from their cocoon, some of them hang around their mother's cobweb for a few days, still eating each other or insects left over from their mother's meals. Before the third molt, however, the young who have triumphed during this highly competitive childhood disperse and build cobwebs of their own. It takes these juveniles another several weeks and three or four more molts to reach full size and sexual maturity.

Good housekeepers might protest, but a human home benefits from a healthy population of house spiders. As these harmless arthropods live out their quiet, unobtrusive lives, they help control the populations of insects—and other arthropods—who seek out human houses. If you must have your ceilings, window casings, and visible corners clean, you can enourage your house spiders to build clean cobwebs by vacuuming away their old dusty ones. Just be sure to warn the resident spider you're coming by tapping its cobweb before you attack. You don't have to love spiders, but some cold winter day when nothing else is happening you might enjoy watching one—and watching one might encourage you to leave these small predators alone.

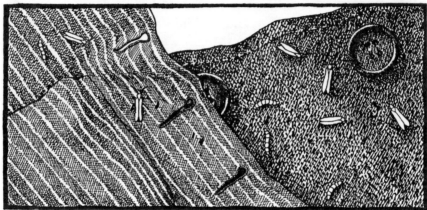

CLOTHES MOTHS

Some of the animals who sneak into our houses are less interested in food than in very specific places to lay their eggs. The clothes moth, for instance, seems to know that human beings hide woolens in drawers and closets. I used to think that all the moths I saw flying around inside my house were threats, but then I started a collection I call "Moths of a Country Home." It consists of all the dead moths I've found in my lighting fixtures and on my windowsills. By looking closely at these moths and comparing each one in turn to a picture of the clothes moth, I have learned that none was searching for my woolens. Like many night-active insects, they were merely attracted to my lights.

The clothes moth, in contrast, is repelled by light. It seeks out dark places—like the drawers and closets where we store our woolens—and responds to light by crawling deeper into folds and crevices.

The females of this species rarely fly at all, and the males fly only far enough to find the females. Despite their fearsome reputation, these moths are only about a quarter inch (.6 cm) long with narrow, satiny, cream-colored wings that spread to about half an inch (1.25 cm). They live for only a few weeks, and, like many other moths, don't feed at all during that time.

It's the young who hatch from eggs laid by these shy, short-lived adults who do the damage. The life cycle of the clothes moth includes the four stages of a complete metamorphosis—egg, larva, pupa, adult—but only the larva is a problem. If you understand the way a clothes moth's life cycle interacts with your own seasonal habits, you should be able to control them without much difficulty.

The clothes moth's timing is perfect. Right when the weather turns warm and you are ready to stuff your wool sweaters and socks into a bottom drawer and store your woolen blankets in a back closet, whatever clothes moths have survived the winter are ready for action. Clothes moths do not enter diapause, but cold temperatures slow down their life cycles, and many spend the entire winter as slowly developing larvae or as inactive pupae. The warming temperatures of spring encourage the pupae to become adults and the larvae to enter pupation, which starts mature clothes moths looking for places to lay their eggs during the early weeks of spring and keeps them looking on into the summer. If you're careless and just stash your woolens out of sight, these clothes moths have it made. Each female merely lays her hundred or so eggs singly or in clusters on the first item she finds, and her young have until you want your woolens again to eat and grow in peace.

Five days to three weeks after the eggs are laid, depending on the temperature, the eggs hatch into a new generation of small white larvae that crave woolens. These larvae are especially fond of woolens that have sweat on them or spilled food, such as tomato juice, milk, beer, coffee, or gravy. They spin silken webs to cover themselves while they're feeding and proceed to chew sizable holes within an inch (2.5 cm) or so of where they hatched.

If these summer larvae are not interrupted, they can reach full size and enter pupation within about six weeks, pupate for another two or three weeks, emerge as adults, and lay another batch of eggs before the weather turns cold again. In a warm laboratory with plenty of food, clothes moths have been known to produce five or six generations a year. If the larvae happen to hatch in a cooler environment—in a box of old clothing stored in a cold basement, for instance—the whole life cycle is slowed down, and one generation might chomp away at the stored goods until they are ready to pupate, which might take as long as four years.

The best way to control clothes moths is to disrupt their life cycle, whatever stage they might be in. Sun is their number one enemy. If you hang the items you plan to store outdoors on a sunny day—and perhaps repeat this process once or twice during the summer—any adults and larvae will drop off the fabric onto the ground, seeking darkness. Beating, brushing, and vacuuming the clothes and blankets will also help dislodge any lurking insects or their inactive eggs or pupae.

If you want to be extra sure that there are no living things among the goods you're planning to store, you can have everything dry-cleaned. Then seal the dry-cleaned items in paper bags or boxes or put them in a cedar closet with mothballs. Both cedar closets and mothballs produce mild fumigants that discourage clothes moths but don't prevent them—especially if they've already gotten a start before you store your goods.

You can also put your winter clothes in cold storage. Clothes moths can't operate below 40°F (4.4°C). Some professional storage companies go to the other extreme, heating goods to 130°F (54°C) before storing them, which is another way of eliminating active insects. And a relatively new approach is to treat woolens with microwaves. Four minutes of microwaves apparently kill all stages of the clothes moth, even the eggs.

Microwaves, heat treatments, cold storage, dry cleaning, moth balls, cedar chests, and modern housekeeping have set clothes moths back a bit, but whatever we do to them, they will still survive as a species. Before we came along with our edible clothing and blankets, the females laid their eggs in old birds' nests, where their larvae found feathers, and dens or burrows, where they found remnants of fur.

I don't like the idea that clothes moths might someday find their way into my house and lay their eggs among my woolens, but when I consider how slipshod my seasonal storage habits are, I wouldn't blame them for trying. When I find myself adding a small, satin-winged moth to my "Moths of a Country Home" collection, I'll know it's time for me to worry.

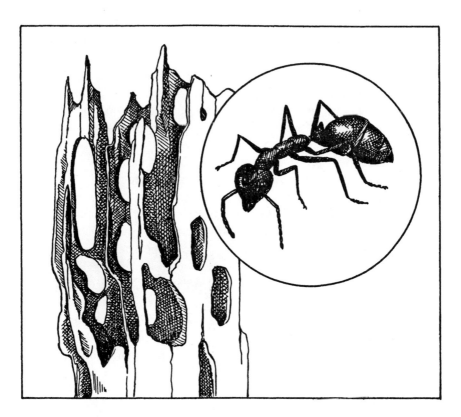

CARPENTER ANTS

When carpenter ants invade human houses, they want more than just a place to lay their eggs. They want to establish a whole colony. A solitary carpenter ant—a shiny black half-inch (1.25-cm) ant you'd never mistake for one of the little black ants who parade through kitchens—is probably just a forager from outdoors looking for something to eat. But if that single ant is a queen, she might be evaluating your structural timbers. Whenever I throw firewood into my basement, I worry that I might be throwing carpenter ants in with it, so I have studied exactly how these social insects live and learned what to look for as indications of their presence.

Carpenter ants are called "carpenters" because they fashion elaborate, galleried homes for themselves inside wood. They don't eat the wood—they merely chew into it and carry each small mouthful to an exit, where it piles up like sawdust. For food they prefer the honeydew of aphids and other plant-sucking insects, plus other sweets, fats, and dead insects.

Outdoors, carpenter ants aren't a problem. They are, in fact, a positive force. They usually carve their galleries into dead or dying trees, thereby contributing to the process of decay. Also, natural predators, such as pileated woodpeckers, keep their populations in check. The large rectangular holes you see in the trunks and limbs of trees mark places where pileated woodpeckers dug in to find carpenter ant galleries and feasted on whatever ants were at home.

For carpenter ants to become a problem inside my house, I would have to offer them two conditions: time and moisture. If I manage to throw a piece of firewood containing a whole colony into my basement, this colony might not need much time to expand beyond the confines of its original log and begin working galleries into a nearby beam, but it would still need moisture. Carpenter ants need a certain amount of moisture in at least a section of their galleries to provide humidity for their developing young. Wood in contact with soil, or wood in contact with other wood that's moist and rotting— such as the flooring under dishwashers or leaking sinks and tubs—is much more vulnerable than high-and-dry beams. I keep a careful watch for the telltale sawdust that would indicate an expanding colony, and if I ever see it, the first thing I will do is eliminate the source of their moisture, which is a condition of the colony's survival.

If I accidentally introduce just a young queen, she will need the same moist conditions plus several uninterrupted years to build her colony to a size where it can begin to do serious damage, which gives me more time to notice the sawdust. A carpenter ant queen has all the equipment and resources to initiate a colony, increase its population, and eventually launch new males and queens to start new colonies, but she has to begin at the beginning.

At the beginning of her own life cycle—which is also her colony's life cycle—she is born into an existing colony and fed by workers until she is ready to mate. At that point she has wings, which she uses only once to take a mating flight away from the parent colony. Winged males are produced along with the colony's queens, and when the time comes for these sexual forms to fly forth and mate, the males emit a substance from glands near their mouths that invites their sisters to fly out of the nest. The timing is such that males and females of the same species all take to the air at the same time, which encourages a certain amount of cross-breeding among neighboring colonies.

After the males and females mate, the males die, but the females break off their wings and start searching for suitable places to start colonies—which is a dangerous time for one of them to happen into your house. Each of these queens digs a small burrow into a likely piece of moist wood and lays her first batch of eggs. When the eggs hatch into larvae, she nourishes them with her own saliva, never leaving the nest to get food for either herself or her young. The individuals that mature from this first brood are quite small, but they are capable of working, which means they can tend the next brood, enlarge the living quarters, and begin foraging for food for the queen and growing colony.

According to one ant researcher, it takes the colony three years to produce the biggest workers, which are the ones we recognize as carpenter ants. By that time the population of the colony has grown to about nine hundred to twelve hundred individuals, but it doesn't produce winged queens and males until the population has reached about two thousand. The original queen, who never mates again after her initial mating flight, is capable of producing eggs for as many as fifteen years. But I hope it doesn't take me that long to notice her colony's sawdust.

Interesting as carpenter ants are, I'm not willing to sacrifice my house to their social development. Perhaps my best defense against them is, ironically, the very act of throwing my firewood into my basement. This annual relocation and burning of wood that might invite a queen to start her colony near my house limits her progress to just one year. Once in my basement—if the queen and her young colony survive my rough handling of their log—they have at most four extra months to establish a more permanent home. But because I keep my closest watch for ants and sawdust near my indoor wood-pile, I am very likely to catch them in the act. A dead or dying tree would have been a better choice than my woodpile or my house because there the queen would have had the extended time she needs to launch her genes.

TERMITES

*T*ermites are often confused with ants, especially carpenter ants, because both live in wood and both are social insects. But these two insects are very different in their relationship to wood and in their social lives.

The species of termite that is the worst threat to houses in the Northeast, the eastern subterranean termite, is, like the carpenter ant, a North American native. Long before European settlers arrived on this continent with building techniques that placed structural timbers dangerously close to moist soil, both of these species lived in old tree stumps and buried wood, with carpenter ants also working higher up into the trunks and branches of dying trees. Termites need to stay close to soil because it provides the high level of humidity their soft, easily desiccated bodies require.

These two species use the wood they live in in entirely different ways. Whereas carpenter ants merely carve their nests into it and forage outside for their sweet, fat, and protein-rich foods, termites find their favorite food right at home. Termites eat wood, or more accurately, cellulose, a complex carbohydrate found in the walls of a plant's cells. Termites can't digest cellulose directly, but numerous cooperative protozoans live in their intestines. These microscopic one-celled animals produce enzymes that convert cellulose to glucose, a simple, easy-to-digest sugar. These intestinal residents, who make the termites' wood eating possible, also offer a possible explanation for the termites' evolution as social insects.

Insect evolutionists theorize that termites evolved not from wasps, as ants did, but from primitive, wood-eating cockroaches, whose intestines also harbored cellulose-digesting microorganisms. These cockroaches had to travel as groups because every time they molted they lost their intestinal residents. The most efficient way for them to renew their supply was to ingest an anal excretion offered by another cockroach who hadn't molted.

Termites, who remain dependent on anal feeding to keep their intestines populated, took this group life a step further, developing castes to divide the labor of a sizable colony. Some individuals became specialized as breeders. These reproductives include the primary reproductives, who, like founding ant queens, stay at home, and younger winged individuals, who, like winged ants, disperse from the colony that raises them. But termite colonies also include other potentially reproductive individuals, who don't have wings but can reproduce if something happens to the primary reproductives—or if they find themselves among a group of nestmates who have become separated from the main colony.

The winged reproductives are capable of starting a new colony by themselves, but their mating system differs from that of the ants. In the spring they fly away from the old colony, drop to the ground a short distance away, and immediately shed their wings. The grounded females then raise their abdomens into the air and, it seems, emit a special scent to help wandering males locate them. Once a male has made contact with a female, the two develop a more enduring relationship than a pair of ants do. Whereas the ants mate immediately and the male then dies, leaving the female to start the colony by herself, the two termites walk off together, the male following the female until she finds a likely place for a nest. They dig into the moist wood together, and only then do they mate. This pair of termites stays together for life, the queen becoming huge as she develops into a full-time egg-laying machine, and the diminutive king remaining at her side to fertilize her eggs as necessary.

Another major difference between ant and termite societies is the sex and status of their young. Ants produce all females, except for an annual crop of males, none of which do anything except eat until they become adults. Termites, in contrast, produce both male and female offspring, and they start participating in the work of the colony while they're still developing. First they become workers that help the king and queen with the tasks of building a bigger nest and feeding a growing population. Later, some of the workers molt into soldiers—termites with the job of defending the colony. These soldiers' mouthparts are specialized for fighting, so they, along with the king, queen, and youngest larvae, must be fed by the workers.

To be sure there will always be enough workers to support the dependents, the colony must regulate the number of individuals who enter each caste. The caste of a developing larva is dictated by hormones that circulate throughout the colony by means of licking and food sharing. The king and queen secrete hormones that prevent the development of reproductives, while the soldiers secrete a hormone that prevents the development of soldiers. Most of the young, therefore, remain workers, but if anything happens to the king, queen, or some of the soldiers, the absence or reduction of the inhibiting hormone encourages some of the young to develop into the caste that is needed to keep the colony balanced and functional.

The most promising approach to termite control today involves interfering with the hormones that regulate the castes. A chemical called methoprene, which is less toxic than other insecticides and also degrades into harmless substances, causes many of the workers to molt into soldiers, creating a feeding problem the remaining workers can't manage. Methoprene also kills the termites' intestinal protozoa, causing the insects to starve to death even if they are surrounded by wood.

Despite this new approach to termite control, I'm convinced that termites will still be around to threaten us as long as we use their food to build our houses. Termites and carpenter ants are not involved in a conspiracy against us. They are merely taking advantage of the alternatives we created when we cleared away the old stumps and dead trees they lived in and substituted the structural timbers of our homes.

BATS

One night during my second or third summer in Vermont, when I was at best still a beginning naturalist, I had just climbed into bed with a good book when something considerably bigger than any of my household anthropods started flying around my bedroom. At first I thought it was a bird, but it didn't thrash around like a trapped bird. It was actually behaving more like a huge moth, but it wasn't flying into my light the way moths do. I was so fascinated by the creature's graceful, noiseless flight that I spent several minutes watching it as it swooped around the room. Then I realized it was a bat.

Despite my newfound interest in the natural world, I panicked. I had heard that bats suck people's blood, get into people's hair, and carry rabies. In a fit of anxiety, I jumped out of bed, ran out the bedroom door, slamming it behind me to trap the dreaded animal inside, and called a neighbor to help me get rid of the bat.

My neighbor merely walked into the bedroom, removed the screen, and let the frustrated animal fly out, informing me that my visitor was a harmless animal named the little brown bat. Once the little brown bat was gone, I got to thinking about it. How could such a small—it was only about 3 inches (7.5 cm) long with a 9- or 10-inch (22.5- or 25-cm) wingspan—silent, and self-controlled animal be the trouble causer my fantasies thought it was? I decided to find out what kind of animals bats really are.

The first thing I learned is that most North American bats are insect eaters. A few feed on nectar, pollen, or fruit, but not one is interested in blood—human or animal. Vampire bats, who do feed on blood, are residents of the American tropics, but they rarely come north of Mexico. And, furthermore, they prefer the blood of cattle. So much for fear number one.

Next, I could find no record of a bat flying into a human being's hair. An old superstition holds that if a bat gets into a woman's hair, she will have an unhappy love affair or, worse yet, die, but the superstition and the fear it expresses are probably based on the bat's habit of swooping close to human heads to snatch up the mosquitoes and gnats that hover there. So much for fear number two.

My only legitimate fear, it turned out, was rabies, but since the bat flying around in my bedroom had no interest in making contact with me, it wouldn't have infected me even if it was carrying the disease. Less than one-half of one percent of bats carry rabies, and they do not become aggressive with it. You yourself must handle the sick animal to become infected, so never pick up a bat—or, if you do, wear thick gloves—and you won't get rabies from one.

With my fears laid to rest, I was free to think about the positive side of bats—their appetite for insects, for instance. The little brown bat has been known to eat seven to eight insects per minute, and one hundred of these small bats ate 42 pounds (19.2 kg) of insects during a summer-long study. Bats, being night fliers, control nocturnal insects much as birds control daytime insects—take away this night shift, and we'd be in serious trouble.

Bats are well adapted for their nocturnal insect chasing. They have evolved a sophisticated sound system called echolocation to help them find insects flying in the dark. A bat emits ultrasonic sounds that bounce off the objects around it and back into its sensitive ears to tell it where everything is. The returning echoes communicate the size, shape, texture, distance, speed, and direction of insects as small as mosquitoes and fruit flies, which are among the little brown bat's preferred foods. The bat pursues the insect that sounds best, catches it in the membrane that stretches between its hind legs and tail—or sweeps it toward its tail membrane with a wing—and eats the insect from this trap.

A bat's wings, which are quite different from a bird's, are designed for maneuverability rather than speed. Elongated finger bones stretch a thin, weblike membrane that is attached along the entire length of the bat's body into flexible "hand-wings." Because these wings aren't stiffened by feathers, bats can change the curvature as they fly, enabling them to vary their speed and maneuver more efficiently at slow speeds. Birds can fly faster than bats, but they can't slow down as fast or twist and turn as acrobatically.

To contend with the absence of insects during winter, little brown bats fatten themselves by eating extra insects during the late weeks of summer and then disappear into caves to hibernate until spring. Among the species of bats that hibernate—some species migrate—the females do something unusual for mammals. They store sperm. Most little brown bats mate in the fall before they enter their winter caves, but the females don't ovulate until spring. The same seasonal cues that invite them to ovulate invite insects to reactivate, so the bats begin their 50- to 60-day gestation just when insect food becomes available again, and the young are born during the summer, when insects are abundant.

You are most likely to hear bats enter your attic in spring, when groups of pregnant females are looking for warm, dark spaces to raise their young. These maternal colonies often return to a favorite roost year after year with the intention of staying until their young have learned to fly in late July or August. They might alarm you with their noise, but when you think about the number of insects they eliminate during their nightly forays, perhaps you can forgive them whatever sounds they make.

If you must get rid of the bats, the most ecological thing to do is to wait until the colony has dispersed at the end of the summer and then close off all points of entrance so they won't be able to establish a roost again the next spring. Meanwhile, if a lone bat ever flies into the part of your house in which you live, you might seize the opportunity to observe this flying mammal in action. It will not attack you. It will merely fly around trying to echolocate an exit. All you have to do—after you've watched the bat long enough to assuage whatever fear you might have had—is open a door or window for it, and the bat will happily find its own way out.

CHIMNEY SWIFTS

Despite my fondest wishes, I doubt if a pair of chimney swifts will ever choose to nest in my chimney. My flue tiles—three separate sets of them feeding up one chimney—are quite small, and their insides, except for the soot and creosote, are quite smooth. What chimney swifts need are chimneys more like hollow trees, which is where this species nested before human chimneys appeared on the scene. They need an opening big enough to accommodate their 12-inch (30-cm) wingspan as they flutter in and out, and they need interior walls that are rough enough for their sharp little claws to hang onto. Chimney swifts perch vertically, pressing their short, spine-tipped tail feathers against the chimney wall to provide additional support.

If you happen to live in a house with just the right kind of chimney—which, of course, only chimney swifts can decide—you might want to spend a summer listening your way through their nesting cycle. Richard B. Fischer spent fourteen summers observing the activities of several families of chimney swifts, and his exhaustive study, *The Breeding Biology of the Chimney Swift*, provides all the cues.

The first sound to listen for, starting in the spring, is the flutter of wings in your chimney. You may have heard or seen chippering flocks of fast-flying, cigar-shaped birds passing overhead two weeks before this first flutter, but some of these birds were headed farther north, and others—the local residents—had to court before they'd be ready to search for a place to nest. The sound of wings fluttering in your chimney is the first indication that a local chimney swift is exploring your chimney as a possible nesting site. The decision won't be made immediately, but if you hear more flutterings as two birds come and go, and especially if they begin to spend the night, you can be fairly sure a mated pair is going to nest with you.

Nest building usually begins during the late weeks of spring, and it involves a characteristic pattern of activity. Chimney swifts tend to build during the afternoon, each partner flying off to find a twig and then returning to the chimney to add it to the nest. A chimney swift obtains its twigs in an unusual way. Because of its specialized, wall-clinging feet, it can't perch on a limb or land on the ground, so it flies up to a dead branch, grabs a brittle twig with both strong feet, and hopes that its momentum will break the twig. If it fails the first time, it circles back and tries again.

When the bird succeeds, it shifts the twig to its beak and flies back to your chimney, where it does another unusual thing. It covers the twig with a gluey saliva that all members of the swift family produce during the nesting season and sticks it directly to the sooty surface it has chosen for its nest. Eventually, after lots of afternoon flights in and out, the two chimney swifts will create a twiggy shelf shaped like a half saucer. If this glued-together construction is not soaked by a hard rain or heated by a summer fire, it's strong enough to support not only the eggs but both adult birds.

The female begins laying her four to five white eggs when the nest is about half done, but you won't hear any significant changes in the pair's activity patterns until she lays her next-to-last egg. That's when the parents start to incubate, which is a full-time job for the next three weeks. The parents take turns, one flying in to relieve the other during the day, and both sleeping on the nest together during the night.

If you have the time and patience, you might want to keep track of the incubation shifts during one day just to see how demanding it is to be a parent bird. The absent partner is off chasing insects and sometimes seems to forget that it has left its mate back in the chimney trapped on the family eggs. The incubating bird does fine for half an hour and doesn't even mind an hour, but if the partner who's feeding remains gone for longer than that, you might hear some fidgetings and flutterings of protest.

The eggs hatch nineteen to twenty-one days after incubation begins, and for the next month you will hear your swifts flying in and out with insects to feed their young. During the first week they deliver meals about every half hour. During the second and third weeks, the interval stretches to about forty-five minutes, and after the third week the schedule becomes somewhat erratic. At that point, the busy parents are trying to find enough insects to feed themselves and also satisfy four or five young who have grown too big to stay in the nest. The young, who can't fly yet, climb out of the nest and cling to the wall nearby, flapping their wings and waiting impatiently for their parents to bring them more food. Each parents flies in about once an hour and offers a pellet of insects and saliva to the young one who makes the most noise.

During the month the young are developing in your chimney, their begging cries will change from weak little squeaks to louder, more assertive bird sounds. These cries will also speed up, until by the time the young are ready to fly, they are no longer begging but chippering like their parents. Young chimney swifts take their first flight when they are thirty days old, so you'll want to be listening for that special event about thirty days after you detected the parents' switch from incubating to feeding their newly hatched young. Be especially attentive during the mornings because that's when young swifts are most likely to flutter out of a chimney for the first time.

For the next week or so, the fledglings will return to your chimney each evening to roost, but then the family group breaks up. You might have a visit from a wandering swift during this end-of-the-summer period, but by early fall all of them are headed south to spend the winter in the upper Amazon Basin in Peru. And you can rest your ears until spring, when you're very likely to hear the flutter of wings in your chimney again as one of last year's swifts explores your chimney to see if it still looks like a good place to raise a family.

V. HOUSEHOLD
NATURAL HISTORY

SPONGES

Along with many of my friends who grew up during the post-World War II era of miracle synthetics, I decided as an adult to surround myself with natural things. It took me a while to figure out what was natural, but once I started paying attention to the origins of various household products, I discovered a wealth of life-improving possibilities. On one of my organic, back-to-nature shopping trips, for instance, I rediscovered the natural sponge.

I'd been using little pink, green, yellow, and blue, perfectly rectangular, totally synthetic sponges for so long that I'd almost forgotten what a natural sponge looks and feels like. The rounded contours, the deep and different-size holes, and the warm tan color attracted both my eyes and my hands, and once I had touched this temptingly squeezable thing, I could hardly wait to get it home and into the bathtub.

What a sponge does in a bathtub is related to what it was doing in the natural world. It was living in the ocean, in a unique association with water. Technically, a sponge is an animal, but it's not much like the other organisms we recognize as animals. It lacks a heart, brain, nerves, lungs, muscles, the power of locomotion, and other attributes that give animals like ourselves a certain physical and behavioral complexity. A sponge is essentially a group of living cells supported by a skeleton, which, in the case of the commercially valuable sponges, is made of a soft, elastic substance called spongin. Fibers of this spongin form a dense mesh, which attracts and retains some of the water that enters a sponge through its numerous pores.

A sponge lives by passing great quantities of water through itself, in some species over a hundred gallons an hour. The water bears small marine organisms and bits of organic detritus, which the sponge eats, and also provides oxygen, which the sponge needs to support its life processes. The major difference between the sponge as it lives in the ocean and the sponge as it performs in the bathtub is that all the living cells, many of which were specialized to keep water moving on through, have been removed, leaving only the inert but highly absorbent skeleton.

While some of the living cells were specialized to draw water into the sponge and others to expel it, still others were specialized to digest food and others to reproduce more sponges. When a sponge is ready to reproduce, it can do either of two things. It can grow branches and lose these genetically identical bits of itself to ocean currents, which might carry them to new locations. Or, if the sponge is big enough and the water temperature is warm enough, it can mix genes with other sponges by producing both sperm and eggs. It keeps its own eggs but sends its sperm out into the ocean on a burst of expelled water, counting on ocean currents to carry it toward another sponge of the right species. This fellow sponge draws the sperm in on an incoming wave to fertilize its waiting eggs. Each fertilized egg develops into a small, free-swimming larva, which passes out of the parent on another burst of outgoing water, and swims around for a day or two until it finds a rock or piece of coral to attach itself to. Once it finds a suitable growing surface, it attaches itself for life, never moving again—unless it gets harvested.

If a sponge is harvested by a technique known as "hooking," which involves a long-handled, three- or four-pronged implement that looks like a garden tool, its life is not necessarily over. The part of the sponge that leaves the ocean, of course, dies, but the small fragment usually left attached to its anchorage will regenerate, achieving its former size in about two years.

This ability to regenerate has caused fierce controversies about how sponges should be harvested. The "hookers" disapprove of the "divers," who harvest sponges from deeper water by cutting them cleanly from the surfaces they grow on. Cutting produces more attractive sponges, but it also eliminates the possibility of regeneration.

When a sponge is hauled out of the ocean, it looks nothing like the sponge I bought at the health food store. It is covered with a slippery black mass of living cells, and it's filled with all kinds of small animals who were living quite happily within the porous body of the sponge. All this living matter must be removed by draining, squeezing, pounding, and scraping, until only the skeleton is left, which is hung up to dry. This skeleton is further cleaned, bleached, and trimmed by buyers and then compressed into bales to be shipped to dealers, who distribute them to stores to be bought by people like me.

Natural sponges began suffering a series of setbacks in 1936, when synthetic sponges became available. The next year a fungus disease wiped out most of the natural sponges growing off the coast of Florida, and then World War II cut off the supply from the Mediterranean. Natural sponges have just never won their way back into modern homes. But with Florida's sponges now growing healthily again, the American sponge industry could regenerate as readily as living sponges do. All the promoters would need to do is let people squeeze natural sponges again, and these soft, durable, highly absorbent—and renewable—natural products would easily sell themselves.

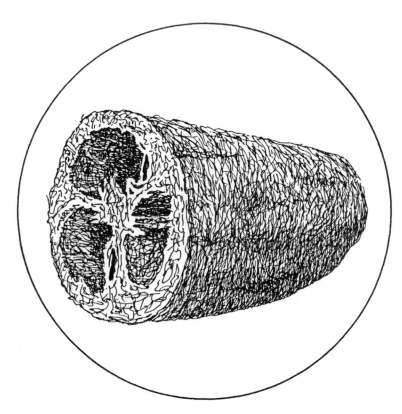

LOOFAHS

Shortly after I had bought myself the natural sponge, a friend introduced me to loofahs. A loofah is a pleasantly scratchy network of straw-colored fibers that looks somewhat like a biscuit of shredded wheat. Used in place of a washcloth, it's supposed to do wonders for the circulation. I bought myself a loofah for the same reason I had bought the sponge: it was a natural product.

This particular natural product, which I had never heard of until my friend told me about hers, has served me exceptionally well. Given the number of times my loofah has been soaked, soaped, rinsed, and dried, a lesser material would have long since disintegrated, but this durable 9-inch (22.5-cm) scrubber still retains its original shape, which resembles a crocodile's snout, and its original texture, which is somewhere between a plastic scouring pad and a sun-dried washcloth.

Unlike my sponge, which was created by an animal, my loofah was created by a plant. This plant, which is an Asian relative of the cucumber, called *louff* by the Arabs, had no interest in producing a scrubber designed to withstand the rigors of the human bathtub. It was merely creating a durable structure to support and disperse its seeds. The crisscrossing strands of the loofah are actually modified veins that hardened as they aged. When the plant was living, they conducted fluids from the fruit's main veins, which run lengthwise just beneath its skin, throughout the rest of the fruit. The walls of the cells that compose these branching and intersecting veins became increasingly woody as the fruit matured, until by the time the seeds were ready to disperse, the veins had formed a persistent and protective skeleton.

If you examine the structure of a loofah, you will see what the plant was up to. Stick a finger into one of the holes that open at the cut end of a loofah, and you will feel a row of chambers where the seeds developed. When the pulp surrounding the seeds dried, the persistent veins formed high-ceilinged tunnels through which the seeds could pass into the outside world. Other veins surround these tunnels, creating a thick mesh around what was the fruit's central core. This durable skeleton enables the seeds to drop out a few at a time over a long period, which improves the plant's chances of reproducing itself.

The fruit that produces the loofah looks like a long fat cucumber, but it's referred to as a gourd. It grows on a climbing vine that uses branching tendrils to ascend 10–15 feet (3–4.5 m) up a tree. As each hanging gourd matures, its bottom falls off, releasing the first few seeds. Some of the seeds fall to the ground directly below the parent plant, but over the extended period of dispersal, winds and storms carry others to a less competitive distance.

The loofah that found its way into my bathtub was harvested, as are all commercial loofahs, at about the time the bottom of the gourd was ready to drop off. The gourd was then soaked in water until the pulp softened, the skin and pulp were washed away, and the seeds were squeezed and shaken out, leaving only the skeleton of hardened veins.

These natural loofahs are valuable for more than just indulgent bathing. The density of their crisscrossing veins enables them to detain oil and dirt while allowing water to pass on through, so they also make excellent filters. These loofah filters can be washed in soap and water and used again, which makes them highly desirable in situations—as on a ship—where a filter must perform over and over again without replacement.

Modern loofah users might be interested to know that the United States experienced a loofah crisis during World War II. The Navy depended on loofah filters for the engines that powered their ships, and at the time of Pearl Harbor, Japan was the sole producer of these filters. Because the Navy needed loofahs, the United States encouraged several Central and South American countries to grow crops of them, but the plants didn't do as well in this part of the world. As soon as the war was over, Japan resumed its monopoly, producing all the top quality industrial loofahs.

When I bought my loofah, I had no idea what it was beyond the fact that it was a natural product. Only when I traced its origins did I learn about the plant that produces these unusual structures and the role they play in the life cycle of that plant. Although I wasn't conscious of what was happening to me at the time, purchasing both a sponge and a loofah—one the product of an ocean-dwelling animal, the other the product of a tree-climbing plant—turned me into an eccentric indoor form of protonaturalist, which, in retrospect, I have named the "bathtub biologist."

COTTON BALLS

If you're feeling frazzled by modern technology, go to your medicine cabinet and find a 100 percent cotton ball. This natural product will take you back to basics. Hold the cotton ball in one hand and pull a few fibers gently forward with the thumb and forefinger of the other, rolling the fibers tightly together as you pull. Without much difficulty you can produce a short length of spun cotton called yarn, and if you are patient enough, you might even succeed in spinning the whole cotton ball into a piece of yarn long enough to do something with.

About seven thousand years ago, a resident of what is now Mexico probably removed the seed hairs from a wild cottonseed, played with them as you have played with your cotton ball, and discovered that these seed hairs could be twisted together and spun into considerable lengths of yarn, which in turn could be woven into fabrics. Archaeologists exploring Mexican caves have discovered the remains of cotton plants and cotton cloth that date back to at least 5000 B.C. The residents of prehistoric Pakistan and Peru apparently discovered spinning and weaving independently and were producing their own cotton fabrics by 3000–2500 B.C.

The seed hairs of the cotton plant lend themselves to spinning because of their construction. Each hair develops from a single cell on the skin of the cottonseed. The cell elongates during the first three weeks after pollination. Then, for the next three weeks, its walls thicken with daily layers of the plant substance called cellulose—the same tough, difficult-to-digest substance that termites eat. By the time the hair cell has developed fully, it is over 90 percent cellulose, which accounts for the strength and durability of cottonseed hairs—and of cotton products.

When the cottonseed itself matures, the elongated and thickened hair cells die, and each cell collapses into a flattened structure that resembles a twisted ribbon. The twist of the flattened cell is what gives a cottonseed hair, known as the cotton fiber, its ability to interlock so strongly with other cotton fibers when they are spun together. Although you can't see the flattened and twisted shape of individual cotton fibers with your naked eye, nor easily isolate just two to see exactly how they interlock, you can certainly feel several of them interlocking when you spin a bit of your cotton ball.

Actually, the fibers that are used to make cotton balls are not even the prime fibers used to spin threads and fabrics. The longest and best fibers, called the lint, are removed from the cottonseeds first. Then the seeds, which are still fuzzy with shorter fibers called linters, are processed again three times until all the linters have been removed. The first-cut linters are cleaned and sterilized to become surgical cotton and cotton balls, while the other two cuttings are used to produce chemical cellulose for film, plastics, and other cellulose-based synthetics.

To compare the linters of your cotton ball to prime lint fibers, pull on a piece of 100 percent cotton thread until it breaks. At the break you will see the tapering ends of some of the individual fibers that were spun together into the strands that constitute the thread. If you pull some of these fibers free, you will see that they are longer versions of the fibers in your cotton ball.

These seed fibers from the cotton plant, which have served the human species for over seven thousand years, probably evolved to protect and perhaps to help disperse wild cottonseeds. They still protect the cottonseeds—so much so that modern cotton growers must remove most of them before planting the seeds to facilitate germination—but they aren't much help in dispersal anymore. When various species of wild cotton became domesticated in different parts of the world, human beings began selecting plants that produced more numerous and longer fibers, until they developed the several types of exceedingly cottony plants that grow in cotton fields today. These cottony plants are desirable for the textile industry that has developed along with the domesticated plants, but the seeds can no longer go much of anywhere without the assistance of human beings.

A cotton plant grows like a small bush. As a wild plant it was a perennial, becoming dormant during the winter and living for several years, but as a domestic plant cotton has become an annual. The seeds are planted anew each spring, and the cotton is harvested at the end of each growing season. The plants flower about two months after the seeds are planted, and the cotton *boll*, which is the flower's enlarged ovary, matures forty-five to sixty days later. The mature boll is only about the size of a golf ball, but it soon cracks open and the cotton fibers fluff out into a bigger, looser structure called a bur. The fluffy cotton in the bur is divided into four or five sections called locks, each containing seven to ten seeds with their seed hairs holding the lock together.

These cottonseed hairs have become important to our species because they can be spun and woven into durable, washable, and, above all, comfortable clothing. Cotton clothing is comfortable primarily because it is absorbent. It wicks perspiration away from the body and allows it to evaporate into the air, which most synthetic fabrics seem to have difficulty doing.

At this point in my life, I have retreated almost completely from synthetics and returned to the comfortable cottons I wore as a child. I wouldn't want to have to make all my clothing by hand from cotton balls, but trying to coax some of those short fibers into lengths of cotton yarn has given me an appreciation of the technology that converts raw cotton into the many cotton products I use today.

WITCH HAZEL

When I consider how hard most drug manufacturers have to advertise to keep us mindful of their particular products, I'm amazed that witch hazel survives. I'd never heard or seen it advertised, but I somehow knew when I was setting up my own household that I should have a bottle of it in my medicine cabinet to treat insect bites, prickly heat, rashes, bruises, and other conditions of the sensitive human skin. This stalwart lotion has endured the onslaught of modern chemicals—and the advertising that sells them—and still asserts quiet spaces for itself in far drugstore corners and on bottom grocery store shelves.

Witch hazel comes from trees, or, more accurately, from shrubs that grow wild in the eastern part of North America. A shrub is a woody plant, usually smaller than a tree, with several slender trunks arising from the same base rather than one central trunk. The witch hazels that grow in my Vermont woods look somewhat like stragglier versions of the lilacs that grow around my house.

It took me a while to find my witch hazels because they look very much like all the other shrubs and saplings that constitute the understory—the smaller trees and shrubs that grow beneath the bigger ones—of the northern hardwood forest. Focusing on the right habitat helped; witch hazels tend to grow near edges and openings rather than in thick, dark woods, and they are often found on brookside banks and near ditches along wooded roadsides.

I was looking for witch hazels not because I was planning to make my own lotion but because they flower in the fall. Many naturalists have celebrated witch hazel—its unusual little fall-blooming flowers and its explosive seeds—in their fall writings, and I wanted to be able to evoke these seasonal phenomena, too. When I finally located a witch hazel, I cut several twigs and brought them indoors to study.

I set some of them in water, as I do with pussy willows in the spring, to see if the flowers would bloom indoors. They did, and I got a good look at the small, yellow, spidery-looking blossoms that are so difficult to spot in the fall woods. Witch hazel flowers don't look much more like what we ordinarily call flowers than pussy willow flowers do. Four narrow, stringlike petals surround four pollen-bearing male parts and one female part, which you need a hand lens to see. Interestingly enough, botanists theorize that the group of trees witch hazel belongs to is transitional between trees like pussy willows, which produce their small, petalless flowers in densely packed structures called catkins, and trees like magnolias, which produce showy, broad-petalled flowers.

If I had left the twig that flowered indoors on the tree, the flowers would have been pollinated, and the bottom of each female flower part would have fattened into a strong little husk with two shiny black seeds inside. In witch hazel, this process takes an entire year. On one of my other twigs I noticed some of the former year's seed husks, which hadn't split open yet. The husks were hard and woody, and I couldn't pry them open with my fingernails. But fall weather apparently provides just the right alternation of temperatures and humidities to split the halves apart and contract them in such a way that they squeeze the two seeds until they fly 25 or 30 feet (7.5 or 9 m) from the parent plant.

The temperatures and humidities in my house must have

been fairly close to those in the woods behind my house because a few days after I had brought the witch hazel twigs indoors, the year-old seed husks split open and shot their seeds across my study. In reading the fall writings of other naturalists, I had learned that Thoreau had had a similar experience, his witch hazel seeds having exploded the night of September 21, 1859.

Just as fall is the season when naturalists write about witch hazel, fall is also the season when manufacturers of the lotion harvest it. They wait until the leaves are off and then cut the whole shrub, chip it, and deliver the chips to one of the two factories in Connecticut where witch hazel is made. The roots are left where they're growing and will produce a new shrub that will reach harvestable size in about ten years. All the witch hazel that is used to produce the annual supply of the lotion is still harvested from the wild, and it remains a renewable resource.

When the chips arrive at the factory, they are chipped again and then steamed to extract the essential ingredients. The healing principle, which has not yet been analyzed chemically, resides in the cambium—the layer of active cells between a tree's bark and wood that produces the tree's conducting tissues and creates a new growth ring every year. This perishable extract is then mixed with alcohol to prevent the growth of molds and bottled to be sold.

I was disappointed that my fresh-cut witch hazel twigs didn't smell like the witch hazel in my bottle. I kept smelling one and then the other, but I couldn't detect even a slight relationship. Then I decided to boil a few twigs—as the North American Indians, who were the first to use witch hazel as a skin lotion, did. Soon my whole kitchen smelled like the inside of a witch hazel bottle as the steam carried the volatile substance around the room. When the boiling twigs cooled, I had a liquid that looked like weak tea and smelled like an alcohol-free version of my bottled lotion. Such homemade witch hazel doesn't last very long, and the tannins that turn the water to the color of tea aren't very good for your skin, but I rubbed a bit onto my chapped hands anyway, and felt satisfied that now I could write about witch hazel as fondly as other naturalists have.

Perhaps the reason witch hazel needs no advertising is that it feels good to the human skin even if no one knows exactly why. The North American Indians, who believed that a fall-blooming shrub should have special powers, taught the early European settlers about it, and word of mouth has done the job since then.

CEDAR CLOSETS

I knew a lot about cedar before I even began to think about trees. One of the houses my family lived in while I was growing up had a cedar closet in the hall right outside my bedroom. Many times I opened the door of that closet just to inhale the strong, clean aroma. I also spent some time looking at the reddish, streaked wood because it seemed like such miraculous stuff to me. What smelled so good to my young nostrils was the bane, so I was told, of the dreaded clothes moth. That cedar closet wasn't exactly an epiphany on a mountain-top or a vision in the wilderness, but we naturalists—especially those of us who grew up in the suburbs—have to accept our formative ex-periences where we found them. For me, a cedar closet outside my bedroom was a starting place.

I encountered cedar again when I moved to Vermont, only this time it was the tree. I was cutting up some wood from a small evergreen that was invading an old pasture, when I caught a whiff of cedar. There, at the center of the log I had just cut, was a thick core of the familiar reddish wood. The freshly cut cedar smelled so good I brought the logs indoors to season in the living room, and later that winter I enjoyed some delicious-smelling fires. Firewood isn't the best use for cedar—it seems almost sinful to burn something that can do so many other things—but a few small logs popping in a hot fire were certainly an indulgence.

The cedar of my childhood cedar closet and of my Vermont pasture is the eastern red cedar, which is a native of the eastern half of the United States. Botanists and foresters will tell you that this North American cedar is not a true cedar, a name they reserve for the cedars of Lebanon and a few other Old World evergreens. But I'm not sure that this distinction matters in the context of the cedar we encounter indoors. Chests, closets, souvenirs, and even the cedar shavings used in pet cages are made from eastern red cedar, so this "cedar" which is technically a juniper, is what most people are talking about when they say "cedar."

The eastern red cedar, or *Juniperus virginiana*, has a narrow, columnar, pointed crown. Unlike most other common evergreens, its needles grow pressed against the twigs, overlapping like scales. The younger needles that grow toward the tips of the twigs are sharp-pointed and prickly to the touch. Instead of cones, the red cedar bears small, dark blue berries covered by a whitish bloom. These berries are a favorite food for several species of birds, including cedar waxwings, who, like chests, closets, and shavings, are named for the eastern red cedar.

Eastern red cedar became popular for building storage chests back in the 1600s, not long after the earliest European settlers discovered an abundance of it growing around their new homes. The first published claim that cedar would protect woolens against insect damage was dated 1682, and scientists have been trying to prove, disprove, or quantify that claim ever since. The only thing that seems safe to say—unless new research being done in India proves otherwise—is that the aroma of cedar in a chest or closet may prevent some moth eggs from hatching and may kill some of the young moth larvae, but it doesn't affect the older larvae, pupae, or adults. It therefore offers no guarantee against the depredations of an established population.

Despite the lack of scientific support for three centuries' worth of folk belief, I still think cedar chests and closets are a good idea. They encourage good storage habits, and they also encourage the use of a natural substance to protect a natural fabric against a natural pest. Furthermore, building yourself a cedar chest or lining one of your existing closets with precut cedar boards will encourage you to observe the natural characteristics of red cedar.

You can buy a closet lining kit at a lumber yard or a home improvement center. The narrow, tongue-in-groove boards in these kits are often covered with a white substance that looks like frost. This "frost" is crystallized cedar oil—an indication that your boards contain an ample supply of this aromatic substance. You can either wipe the crystals off, or they will disappear by themselves after the boards have been out in the air for a while.

Most of the boards will be dark pinkish red—the color of red cedar's heartwood. Heartwood is the core of dead wood at the center of a tree, and in red cedar it occupies almost the whole trunk. It is here that the cedar oil—designed to protect the slow-growing, long-lived tree from insect attacks and early decay—is concentrated. The lighter, cream-colored streaks on some of the boards, which make cedar so attractive on closet walls, or as brightly polished little souvenirs, are bits of sapwood. Sapwood is younger than heartwood and in the living tree still participates in life processes. In red cedar, it is just a narrow ring around the predominant heartwood.

Many of the boards in a closet lining kit will have knots, indicating places where branches were growing from the tree trunk. They may look like imperfections, but the more knots the better for your closet lining because knots contain even more oil than the straight-grained heartwood. You will become thoroughly accustomed, perhaps even addicted, to the aroma of this oil while you're cutting and nailing your cedar lining in place. Should the aroma lessen over the years, you can revive it by rubbing the walls of your closet with fine sandpaper or steel wool.

I haven't installed a cedar closet in my Vermont farmhouse yet, but I've already bought a kit, and the construction is on my list of rainy-weekend or dead-of-winter indoor projects. In addition to providing me with a storage space that might discourage clothes moths, this closet will also offer me a place to stick my head into once in a while for old times' sake. I'm glad to have finally made the acquaintance of the living tree that produces such attractive, aromatic wood, but my real interest in cedar is derived from that suburban cedar closet.

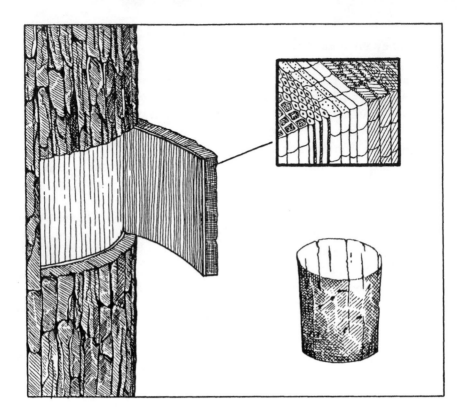

CORKS

I save corks. Every time I pull one out of a wine bottle I think I really ought to throw this one away, but then some Yankee sense that I might someday find a good use for old corks takes over, and I add it to my collection. Perhaps the reason I'm so attached to corks is that they, too, come from the natural world, from a tree that grows near the Mediterranean.

This tree, which is called the cork oak, needs extra protection against the hot, sand-blasting winds that blow across the Mediterranean from the deserts of North Africa, so it has evolved exceptionally thick layers of cork, the substance produced by a tree's outermost ring of living tissue to insulate and protect its sensitive inner tissues.

This outermost ring of living tissue is called the *cork cambium* to differentiate it from the cambium, which is deeper inside the tree. While the cambium produces cells that become the tree's conducting tissues and eventually its wood, the cork cambium produces cells that become the tree's protective tissues and eventually its outer bark.

Cork cells spend most of their brief lives thickening their walls with a waterproof, airproof substance called suberin. By the time the cork cells die, their walls are impermeable, and a bit of air is trapped inside. These impermeable, air-filled cells are bonded to one another by natural resins, and all of them together constitute a continuous, impermeable sheath around the tree. Eventually, the outermost layer of cork and whatever other tissues were caught outside this impermeable barrier darken and crack, split, flake, or peel into the characteristic pattern of the tree's bark.

Cork oak differs from other trees in that it produces thicker layers of purer cork. There are no fibers or other remnants of noncork tissues mixed in with it, so it is a completely and uniformly cellular material. Robert Hooke, an English scientist who pioneered in the use of microscopes, discovered this fact in 1665. In describing what he saw when he looked at a piece of cork under his microscope, he used the word "cells," introducing the word—and the concept—that has become the basis of modern biology. Two and a half centuries later, scientists turned to cork again to demonstrate Lord Kelvin's hypothesis that cells must take a fourteen-sided shape, called a *tetrakaidecahedron*, to occupy space as uniform bodies of minimal surface without leaving any gaps. That's exactly what cork cells do.

When you look at a wine cork or a tapered stopper you can buy at a hardware store, you can't see the tetrakaidecahedral structure with your naked eye, but you can observe the effects of the efficiently arranged, air-filled cells. The air in the cells is what allows a cork to be forced into a space that seems too small for it—like the neck of a wine bottle. As the cork is pushed into the bottle, the air is compressed. It can't escape from the cells unless the cell walls are ruptured, so the cork is also resilient. When you open a bottle of wine, even years after it was sealed, the cork expands to its former size and is difficult to force back in. While you're pushing and pulling the cork into and out of the wine bottle, you will also notice that it doesn't slide easily. That's because the cut edges of the microscopic cork cells act like little suction cups, forming partial vacuums between themselves and the smooth glass.

Despite all the pits and holes in a typical piece of cork, a wine cork is impermeable to air and moisture. To understand why, examine the top of your cork. You will see some dark lines that look like cracks or channels, and at right angles to these lines, some very faint lines that seem to be embedded in the cork. These faint lines are growth rings, indicating annual increments of cork—and the position of the wine cork when it was still a part of the tree.

Now, if you hold the cork as if it were still growing on the tree, you will see that the dark channels cut right through the growth rings. These dark channels are lenticels—the pores that allowed the tree's living tissues to breathe through the otherwise impermeable cork. But air can move only through these lenticels. It can't move at right angles to them, which is how the cork is cut. Therefore, a wine cork is impermeable from top to bottom, or, when it's sealing wine inside the bottle, from bottom to top—as demonstrated by the pop of escaping air that accompanies removal of the cork.

When the cork oak tree is about twenty years old, it has produced enough cork to spare some for human uses. This first stripping is called virgin cork, and it is too rough and cracked to be used for wine corks or bottle stoppers. Most of it is ground and then baked in molds until natural resins bond the granules into solid blocks called corkboard, which is used for insulation.

After the virgin cork is removed, a new cork cambium forms and starts producing more cork. This new cork cambium produces cork faster than the first one did, and after only ten years, another thick layer of cork can be harvested. This second harvest yields a better quality cork, but it's still not good enough to become wine corks. It's only the third and subsequent harvests, which are taken every eight to ten years until the tree is one hundred fifty to two hundred years old, that produce the high quality cork that is cut into wine corks. Bulletin boards are made from scraps of this high quality cork, which are ground into small granules and glued into what is called composition cork.

I guess when I think about what cork is and where it comes from, the corks I save aren't that much crazier than the nutshells, bird feathers, snakeskins, and dead insects I pick up to add to my collection of natural curiosities. A wine cork is just as natural as these other things—it's just been cut and shaped to a human purpose.

CHOPSTICKS

Whenever a new Chinese restaurant opens, I have to eat a meal there to see if it approaches my standards for what a Chinese restaurant should be. Once I walked into one the first day it was in business and was presented with a brand new pair of chopsticks. While I was waiting for my dinner to arrive, I found myself examining these familiar utensils as if I had never seen them before.

Elegant chopsticks might be made of ivory, silver, or enameled wood, cheaper ones might be made of plastic, and Japanese restaurant chopsticks are sometimes made of a lightweight wood designed to be thrown away after just one use. But ordinary chopsticks of the type that are likely to be given away free at the opening of a Chinese restaurant are made of bamboo.

To be sure your chopsticks are bamboo, look for certain identifying characteristics. They are usually a little over 10 inches

(25 cm) long, round at one end, and square at the other. When you hold the chopsticks horizontally, as if you are about to pick up a morsel of food, you will see parallel streaks running their entire length. If you point the round ends toward you—or the square ends for that matter—you will see some of these streaks in cross section. From this angle they look like exactly what they are: vessels, which in the living bamboo plant transported water and mineral nutrients from the soil toward the leaves, and food from the leaves back down to the underground storage organs and roots. It is characteristic of bamboos to have their conducting vessels scattered throughout their stem tissue rather than arranged in rings as they are in trees.

Because bamboos are tall, woody, and perennial, it's easy to think of them as skinny trees, but they are in fact very different plants. A tree begins as a small sapling and grows a little taller and a little thicker each year. The stem, or trunk, is solid, consisting of heartwood and sapwood, with concentric growth rings representing each year the tree has been alive. A bamboo stem, in contrast, is hollow, and it achieves all its growth in one spectacular season.

The shoot emerges from the ground as thick in diameter as it's ever going to be and bolts to its full height at a rate that in some species has been recorded at almost 4 feet (1.2 m) per day. While a tree spends each growing season manufacturing food for its own needs, a bamboo stem spends the fifteen or so seasons it might live manufacturing food for new stems that arise from the same root system. These new stems are too busy growing during their first season to make food for themselves.

Despite the treelike appearance of bamboos, these hollow-stemmed plants are actually more closely related to grasses. Like lawn grasses and the grain crops that grow in farmers' fields, bamboos have jointed stems (called culms), narrow, pointed leaves, and flowers that grow in spikelets at the top of the plant. But unlike the lawn grasses and grains, they do not die back at the end of the growing season, and they don't flower every year.

Some bamboos flower sporadically, but many flower cyclically, demonstrating one of the most unusual life cycles in the entire plant kingdom. When one of these bamboo seeds germinates, it produces an aboveground plant consisting of vertical stems and an underground support system consisting of horizontal stems and roots. Every year, these underground stems, which are thick food storage organs called rhizomes, send up more bamboo shoots to bear more leaves and manufacture more food. Finally, after the bamboo plant has been growing and producing genetically identical stems for any-where from twenty to one hundred years, depending on the species, all the aboveground stems produce their flowers and die.

Because all the members of any one of these species are genetically programmed to flower at about the same time wherever they might be growing, different species of bamboo are subject to sudden disappearances. Eventually the seeds that result from the pollination of the flowers will produce a new generation of long-lived plants, but during the five to ten years it takes for the new plants to grow a number of stems, there's likely to be a shortage of any species that has recently flowered. The giant pandas of China, whose limited habitat has made them dependent on just two species of bamboo, have been in serious trouble since the mid-1980s because both of their species have just bloomed, and the plants are dying off faster than wildlife researchers can find the pandas alternative sources of food.

By the time my Chinese dinner arrived, I had become so obsessed with bamboo that I searched madly through the stir-fried vegetables in hopes of finding a bamboo shoot to look at, too. The Chinese not only use bamboo to make chopsticks—and tea strainers, baskets, mats, furniture, rakes, fishing rods, musical instruments, toys, backscratchers, water pipes, and scaffolding—they also eat it in the form of its young shoots. These shoots are harvested just as they break through the soil, while they're still white and tender. The outer covering of the shoot is removed, and the tender flesh is boiled before it's eaten. The bamboo shoots served at Chinese restaurants—or that you can buy canned at a grocery store—are cut into small pieces, so you can't see the whole shoot, but you can see the fibrous conducting vessels that were developing in what was to become the bamboo stem.

I was beginning to feel an almost mystical fusion with bamboo as I pinched a piece of shoot between the tips of my chopsticks and held it up to the light to see how its internal structure might resemble the internal structure of the chopsticks. The Chinese are accustomed to seeing art, beauty, and metaphorical messages in bamboo, but I'm not sure they have many of their visions in restaurants. I decided I'd better eat the shoot and metabolize some of bamboo's wonders before the management invited me to leave.

BROOMS

Housecleaning requires a degree of commitment I don't often feel. When I do set myself to indoor chores, I keep myself going by thinking about such inspiring subjects as the history of housekeeping and the domestic technology I am heir to. One weekend's efforts were sustained in large part by a meditation on brooms, thanks to an old-fashioned, handcrafted broom I happened to have bought from a craftsperson who was displaying his wares at a shopping mall.

I decided to use this old-fashioned broom on some of my rugless floors to see what it might have felt like to be an old-fashioned housekeeper, and I was surprised to discover that there was a certain pleasure in sweeping with it. The broom head is narrower than the modern broom's, but it's a good 7 inches (17.5 cm) longer, and it is sewn flat about 3 inches (7.5 cm) higher, which makes the numerous sweeping tips more flexible. Furthermore, because there are no closely wrapped wires or labels to hide details of construction, I can see exactly how this common domestic tool was made.

The idea of a broom probably began with branches used to sweep the floors of cave dwellings and primitive huts. The first recorded broom, called a besom, consisted of a bundle of twigs tied to a handle. As recently as our own colonial era, settlers tied branches of hemlock, birch, or elm together to create rough but functional tools for sweeping.

In southern Europe, however, the broom was following a different line of development. Sorghum, a tall annual grass that looks somewhat like corn, had found its way from Africa to countries around the Mediterranean. Sorghum produces a stiffly branching flower head at the top of its stalk, and in some plants this flower head resembles a long, slender brush. People began selecting the plants with the longest, most brushlike flower heads to make whisk brooms, and eventually developed the variety of sorghum called broomcorn, which now produces 12- to 24-inch (30- to 60-inch) flower heads suitable for making full-size brooms. While other sorghums are grown for sweet syrup, grain, and fodder, broomcorn, which was first cultivated in Italy during the late 1500s and officially described in 1658, is grown for the sole purpose of making brooms.

Benjamin Franklin is credited with introducing broomcorn to this continent, having supposedly planted a seed he'd plucked from an English lady friend's whisk broom. This event occurred sometime around 1725, and by the early 1800s locally grown broomcorn supported a sizable broom-making industry in Hadley, Massachusetts. As settlers moved west, they took broomcorn with them, and gradually this drought-resistant crop found agricultural niches for itself in the semi-arid regions of Oklahoma, Colorado, New Mexico, Kansas, and Texas. Only one state east of the Mississippi—Illinois—continued to produce broomcorn, mostly for seed. Because growing broomcorn is labor-intensive, requiring hand harvesting and manual processing, it is not grown as a commercial crop much anymore, but anyone can still grow it as a garden crop anywhere that corn can be grown.

Growing broomcorn is, in fact, much like growing corn. Both can be planted in hills or rows after the danger of frost has passed and need only to be weeded during the growing season. The two plants even look alike until they begin to flower. Whereas corn produces its flowers in structures that become the cobs, broomcorn produces its flowers in structures that become the broom straws. The flowers grow at the tips of slender, parallel branches that emerge from a central stalk like a tall, stiff plume. This flower cluster, which is called a brush, is what makes broomcorn so perfect for sweeping. The fine, tapering branches that increase in number toward the tip of the brush, hold dust and other small bits of household detritus much more effectively than the twigs and branches of trees.

My old-fashioned broom shows exactly how the individual broomcorn brushes are combined to make an efficient sweeping tool. The broom maker left the brushes intact, each with a 4- to 5-inch (10- to 12.5- cm) section of the stalk that supported it in the field. He then bound twenty of these stalks tightly to the handle of the broom. Finally, he bound the flaring brushes toward the top and sewed them flat lower down to make a strong and tidy tool. This particular broom maker even left a few seeds attached, which show how the flowers were arranged on the branched and tapering tips of each brush.

Commercial broomcorn is usually harvested before the seeds ripen. If the seeds are allowed to mature, the brushes sometimes splay, bend, become brittle, or turn red. While broomcorn is harvested with 6–10 inches (15–25 cm) of stalk still attached to the brush, modern broom manufacturers cut the brushes free from their stalks, sort and grade the individual broom straws, dye them green, and then bleach them to create a uniform product. If you pull a green broom straw out of a mass-produced broom, however, you will still be able to see the tapering structure that characterized each component of the broomcorn's tall brush.

Sweeping floors will never be my favorite activity, but something about being able to see the structure of the plant that inspired the broom makes my sweeping seem more connected to the natural world. There's a certain logic to using a plant to do a job it seems perfectly designed to do, and added to that there is the ingenuity that converted the plant, without changing it much, into a perfectly designed domestic tool. Somehow, though, my old-fashioned, handcrafted broom has made brooms seem less like tools than functional pieces of organic art.

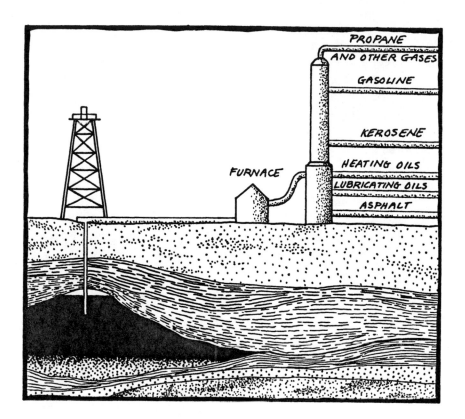

OIL

I tend to divide the world around me into two categories: things that are alive and things that are not alive. Plants and animals are obviously alive, but I also include among my "alives" things that were once living, such as the logs I throw into my basement every winter to burn in my combination wood- and oil-burning furnace, and things that will someday be alive, such as the seeds I buy to plant in my garden.

In my oversimplified, dualistic perception of the world, my "not-alives" include rocks, minerals, metals, and synthetics. I used to think oil was nonliving, in the same category with aluminum, plastic, and other industrial substances that are polluting the earth. After all, it's intimately associated with my car, trucks, jets, and other machines and inventions that complicate my fuzzy-headed life. I'm always pleased when I discover that my assumptions have been wrong, but when the oil crisis of the 1970s forced me to learn about oil, I had to revise my entire world view.

Oil isn't alive in the sense that left to its own devices it might germinate or start breathing or suddenly reproduce itself. But it was once alive—as alive as the trees that have become the logs in my woodpile—only it's been dead longer and subjected to conditions my woodpile will never see. The current theory is that oil is of marine or lacustrine (lake) origin. Millions of years ago when much of the earth was covered by shallow, shifting seas and vast inland lakes, the plants and animals that inhabited these seas lived and died and settled toward the bottom. Their remains were rapidly buried by muds and sands that prevented them from being scavenged by other organisms and also from being completely decomposed.

In time, the hydrogen and carbon atoms in the fats, waxes, resins, and other components of these plants and animals were altered into the many different hydrocarbons we find in oil today. Similar hydrocarbons are also present in coal, which isn't surprising because coal also had an organic origin. The lush vegetation of swamps that flourished during warmer times became coal when it was buried and compressed under heavy layers of sediment.

As the earth's crust has arched, tilted, broken, and, in some areas, been penetrated from below by movements of salt from deeply buried evaporated seas, pockets of oil have been trapped by impermeable barriers. Geologists are doing their best to piece together the earth's history in order to determine the exact locations of the ancient bodies of water and the exact movements of the earth's crust. Only with this background information can oil prospectors calculate where new deposits of oil might be, which they can prove only by drilling expensive wells.

The days of striking oil at 69½ feet (20.85 m), as Edwin L. Drake did near Titusville, Pennsylvania, in 1859, or of gushers like the one that came in at Spindletop, Texas, in 1901, are long gone. Today an oil prospector might have to drill more than 25,000 feet (7500 m) to find oil, which would cost him or her more than $4,000,000. With a barrel (42 gallons or 159.6 liters) of oil selling for $30.00 or less, the prospector would need to strike an enormous reserve to make such an investment worthwhile.

Oil as it emerges from a well is called crude oil, and it can vary from a thin liquid to a thick, black, gummy substance. Before this crude oil can be used by consumers, it must be piped to a refinery, where it is separated into several different components, called fractions, by a process called distillation. The crude oil is heated and then pumped into a tall tower, where the lightest, most volatile fractions emerge from the top as gases, and the heaviest, least volatile fractions accumulate at the bottom as semisolids.

Between these two extremes, all the familiar petroleum products are formed. Toward the bottom, the tower produces asphalts and heavy oil used for lubrication. At the next level up, it produces somewhat lighter oils used for heating—like the fuel oil my furnace switches over to if I let it run out of wood. At the next level, it produces kerosene, which I can use in lamps and space heaters, then gasoline for my car, and finally, at the top, gaseous propane for my kitchen range.

Learning that oil is organic has made me respect it more, but no more willing to be completely dependent on it. If I could drill an oil well in my own back yard and construct a small refinery about the size of a sugarhouse to turn my Vermont crude into lubricants, heating oil, kerosene, gasoline, and propane, I might feel more comfortable with its role in my life. But given the distances today's oil must travel and the very real possibility that people and politics might keep it from making the trip, I think I'm better off throwing wood into my basement—and using the time to reconstruct my world view.

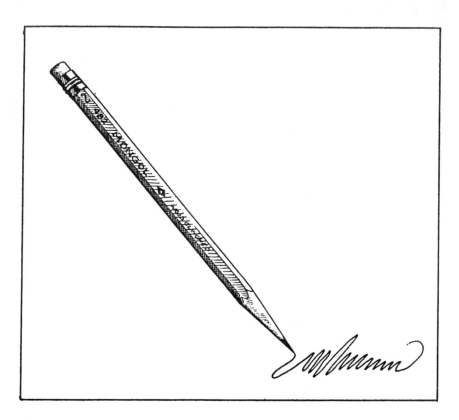

PENCILS

I sharpened a lot of pencils while I was working on this book. A freshly sharpened pencil gave me the right attitude toward revising, and the act of sharpening offered me convenient breaks from my labor. I found myself not only sharpening my pencils but examining their lead, rubbing dirty smudges of it between my fingertips, and even looking at the different marks different pencils made through my hand lens. At one point I dropped the chapter I was revising and addressed my full attention to lead.

The first thing I learned is that pencil "lead" is not really lead. It is called "lead" only because until 1779 everyone thought it was merely a darker form of the familiar metal. In that year a Swedish chemist named Karl Wilhelm Scheele proved that the substance called pencil lead was actually a form of carbon. Eventually it was given a new name—graphite, from the Greek word *graphein*, meaning "to write."

Although graphite is technically a mineral, defined as an inorganic substance, it—like oil—was once living. It is derived from plant matter that has been metamorphosed by heat and pressure into almost pure carbon. It's related to both coal and diamonds, coal having been subjected to less heat and pressure, diamonds to more.

In graphite, the carbon atoms are arranged in layers, the atoms right beside each other closely linked but those above and below only loosely linked. This internal structure makes graphite tend to slip apart in thin, easily broken sheets. When you rub it, it will cover your fingers with dark little pieces of itself, which at the tip of a nicely sharpened pencil would form a dark line on a piece of paper.

Human beings had devised lots of implements to draw and write with before they discovered graphite. They used chalk, bits of burned wood, brushes made from plants or animal fur, reeds, and quills. The Egyptians, Greeks, and Romans used pieces of the metal lead to draw lines, and during the 1300s artists used thin rods of lead for artwork. But it wasn't until 1564, when a deposit of almost pure graphite, called *plumbago*, or "that which acts like lead," was unearthed in Borrowdale, England, that pencils as we know them became a possibility.

At first people just used chunks of this pure graphite, but that got their fingers dirty. So someone decided to wrap twine around long, narrow pieces of graphite to create a primitive pencil. Next came the idea of encasing the graphite in wood, and the pencil began to take on its modern form. But these early pencils were still clumsy, messy tools that produced thick, dark, easily smeared lines. Finally, in 1795, a French inventor named Nicholas Jacques Conte, who was commissioned by Napoleon to come up with a French version of the pencil, thought of mixing moist clay with powered graphite and baking this mixture into pencil leads. He discovered that he could create leads that would write lighter, thinner lines by adding more clay and darker, thicker lines by adding less clay.

If you'd like to explore some of the effects that are now possible with the varying proportions of clay offered by modern pencil manufacturers, buy yourself several pencils with different numbers on them. The standard number 2 is a mixture of about two-thirds graphite and one-third clay. If you like darker, broader lines, you can use a number 1, which includes less clay. For lighter, finer lines, try a number 3 or 4, which include more clay. You may also notice numbers 2½, 2⁵⁄₁₀, and 2.5, all of which offer the same grade of lead. The numbers are different because when the Eagle Pencil Company came out with a number 2½, the numerals became part of their trademark. Other pencil companies that wanted to produce the same intermediate grade of pencil, which consumers seemed to like, were prevented by law from calling their pencils 2½, so they used 2⁵⁄₁₀ and 2.5 instead.

If you want to experiment with the extremes that are available today, buy some art and mechanical drawing pencils. Art pencils rated 4B or higher produce lines darker than the number 1 writing pencil. A 6B makes soft, bold lines that looked like black crayon through my hand lens. For the other extreme, look for mechanical drawing or drafting pencils rated 4H and higher. A 6H makes exceedingly fine lines that look like spider silk next to the strokes of the 6B.

Interestingly enough, none other than Henry David Thoreau was one of the early innovators who helped to develop the kinds of pencils that are now available in this country. His father was a pencil maker, and Thoreau went to work for him when he graduated from Harvard. Because young Thoreau was dissatisfied with the family pencil, he studied European pencil-making techniques to learn how other countries made their superior products. He discovered that the Germans were adding a very special clay from just one mine in Bavaria, found a supply of this clay, and added it to his family's formula. Then he decided that the graphite should be ground finer, and all the grinding that ensued, one scholar suggests, may have exacerbated the tuberculosis that killed both Thoreau and one of his sisters at an early age.

When I'm struggling with something as frustrating as revising what I've written, I'm easily distracted. Most of my distractions make me feel guilty, but my study of pencils—especially when I learned that Thoreau, too, had taken a serious interest in them— seemed legitimate respite. After all, if I'm going to continue to write, I need lots of pencils, and sharpening all those new pencils I bought gave me an excellent attitude toward getting back to work.

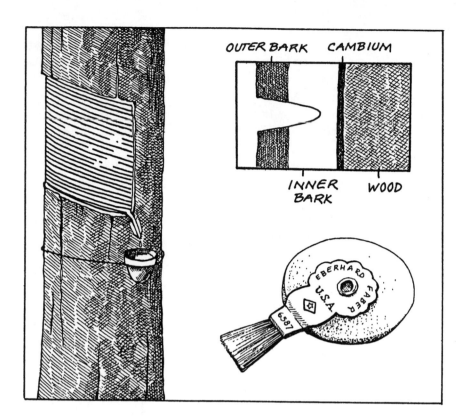

OUTER BARK   CAMBIUM

INNER BARK   WOOD

ERASERS

*I*t's hard to think about pencils without thinking about erasers, and erasers, of course, are made of rubber. Rubber, in fact, got its English name from its ability to rub pencil marks off a piece of paper. Whereas the South American natives, who happened to discover that this versatile substance could be extracted from trees, called it words that sounded like *caoutchouc,* meaning "weeping wood," the English chemist Joseph Priestley, who happened to rub a small piece of it across some pencil marks, called it rubber. Priestley's historic erasure occurred in 1770, about thirty-five years after rubber arrived in Europe.

      The rubber that found its way to Europe was a soft, gummy substance that became runny and sticky when hot and stiff and brittle when cold. It came almost straight from the South American trees. Several species of New World trees produce enough latex—the milky white fluid that other plants such as milkweeds, dandelions, and gold-enrods also produce—to be good rubber trees, but only one, *Hevea brasiliensis*, attracted the attention of Europeans.

This tree grows wild in the Amazon Basin, where natives searched through the dense tropical rain forest for a sizable specimen, gashed its bark, attached a container to catch the latex that dripped from the wound, and returned later to collect it. Because the fresh latex was watery and perishable as it came from the tree, the native collectors treated it over a smoking fire to make it more manageable. They dipped a stick into the liquid they had gathered and rotated the stick above the fire until the water evaporated and the latex solidified. Then they dipped the stick again, gradually building up a large, well-smoked ball of rubber that could be transported out of the rain forest.

The South American natives—and the Central Americans to the north, who extracted their latex from different plants—had learned that rubber balls would bounce, that rubber vessels would hold liquids, and that rubber footwear was waterproof before Europe-an explorers arrived in their part of the world. The European scien-tists who obtained specimens of this New World product from the explorers were fascinated by its properties and set to work on trying to understand why it behaved as it did and how else it could be used.

About fifty years after Joseph Priestley discovered the first "rubber," which replaced moistened bread as an eraser, Charles Macintosh created the first raincoat. And Macintosh's partner, Thomas Hancock, in experimenting with thin cuts of rubber tubing, invented the ancestor of another product I find almost as necessary as erasers—the rubber band. But most of our modern uses of rubber awaited an American innovation, Charles Goodyear's discovery of vulcanization. Vulcanization is a process by which raw rubber is mixed with sulfur and heated to become more durable and elastic than the natural product.

After 1844, when Goodyear patented his vulcanization process, the interest in South American rubber increased signifi-cantly. Many wild rubber trees were ruined by careless gashes cut so deep they damaged the cambium, that sensitive living layer that pro-duces the tree's conducting tissues. And the distances the natives had to travel to find new trees and recover their latex made the whole operation unwieldy. When in 1876 an Englishman named Henry Wickham exported enough *Hevea brasiliensis* seeds to start rubber plantations in other parts of the world, the South American rubber business declined and eventually collapsed.

When the cultivated rubber trees started to grow in their new environments—on Sri Lanka and other islands in tropical Southeast Asia—researchers paid close attention to exactly how the trees produced their latex. The latex, which is totally unrelated to the tree's sap—the fluid that transports nutrients throughout the tree—is located in tubes just beneath the bark. The trick is to cut deep enough to sever the latex tubes without damaging the cambium. If the cut is sloped at a 30° angle, the latex will run downward and gather in a cup attached at the bottom.

The latex runs best during the morning, and each cut will produce for about two hours. Then the latex dries and hardens, as if to heal the wound. After a day's rest, another narrow cut can be made right below the old one. It takes a skilled worker about a month to cut about one inch (2.5 cm) down the tree and three years to work from the top cut, which is about 3 feet (.9 m) above the ground, to ground level. Then the worker moves around to the other side of the tree and works on that side while the bark on the first side is regenerating. Under this sort of regimen, rubber is a renewable resource, with each tree in production from the time it's six or seven years old until it's about forty.

Since the early days of rubbing a raw, unvulcanized piece of this unique natural substance over pencil marks, rubber has changed considerably. Much of it, including what's used in making erasers, is now produced synthetically, but about a third that is used in general manufacturing still comes from trees. While trees and chemists will continue to produce the raw material from which all kinds of rubber products can be made, I find myself wondering if erasers have a future. I still use them—and own an interesting assortment of the different kinds I've accumulated over the years—but self-correcting typewriters and word processors threaten to make the rubbing of rubber across writing an antiquated act.

DUST

I'm not much of a housekeeper, as I have already indicated, but when I'm expecting guests I usually try to clean at least the surfaces of all the rooms they'll see. One time I got sidetracked. As I moved from room to room surveying the job to be done, I was amazed at the amount of dust that had accumulated everywhere I looked. Under the living room sofa I found dust balls bigger than the palm of my hand. My rarely used dining room table was covered with a fine gray film so thick I could write in it, and the windowsills in the hallway were coated with a darker, grittier substance. By the time I started upstairs I was noticing dust in the heating grates, dust on the moldings, and even dust clinging to old spider webs.

Instead of attacking the dust as I should have, I plopped down in my comfortable reading chair to think about it. Up flew more dust, which I could see as fine particles floating in the sunbeams that were shining in the window. Rationalizing that I was a naturalist, not a chambermaid, I decided to examine my own house dust to see exactly what it was made of. If I knew what it was, I thought, maybe I could prevent it at its source rather than having to vacuum, sweep, and wipe it up so often.

But dust, even house dust, is not a simple substance. Its composition varies from house to house, from neighborhood to neighborhood, from life-style to life-style. Much of it enters from outdoors, and the dust that floats around in the atmosphere differs with the environment, the time of day, the season, the altitude, and other variables. Beyond household and local dusts are such invisible but pervasive particles as volcanic and meteoric dusts, and finally something which we will never have to vacuum out of our household corners—cosmic dust.

My own household dust reflects the way I live. Much of it is fine particles of dirt that are tracked indoors on my hiking boots. Because I live on a dirt road, a lot of these fine particles don't even need my boots to carry them indoors. In the summer, many of my windows are wide open, and every car that passes creates a cloud of dust, much of which settles throughout my house. In winter, my wood-burning furnace and fireplace add fine bits of ash to my house dust. The clothes I wear add small fibers of wool and cotton, and I was surprised to see how many of the dust balls I collected were held together by strands of my own hair. Once I started paying attention to the composition of my household dust, I saw myself creating it in everything I did. Even my pencil sharpenings and eraser rubbings are components of my personal dust.

A country house located on a dirt road and heated with wood might seem dustier on a day-to-day basis than a city apartment, but there's a significant difference in both the content and overall quantity of rural and urban dust. Urban dust includes many more particles that are byproducts of industrial processes. More cars also contribute more byproducts of internal combustion. In some cities, the particulate counts—the number of particles floating in a measured volume of air—drop considerably on Sundays, when factories are closed and the twice-a-day rush hours are not in operation. Typical particulate counts, which are measured per cubic centimeter (cc), vary from 5–35 particles per cc at heights over 15,000 meters (about 50,000 feet) to 500 in relatively pure air, and 50,000 in polluted air.

Some dusts are directly related to human activities, but others are independent of what we do. For instance, when a volcano erupts, fine volcanic dust drifts into the atmosphere, where some of it stays for long periods before settling back to earth. Krakatoa, which erupted in 1883, dispersed about 5 cubic miles (21 cubic kilometers) of dust into the atmosphere, and much of it continued to float around for three years.

Meteoric dust is even less observable than volcanic dust. When a volcano erupts, we at least hear about it, and if we live close by we can see the dust on cars and buildings. But we have to be lucky to notice a shooting star, which tells us that a meteoric particle has just entered our atmosphere. What's left of it will eventually settle to the earth's surface as dust, but chances are we'll never see this meteoric dust, or even if we do, we probably won't perceive it as anything different from the dusts that originated on earth.

Cosmic dust is almost an abstraction. It exists, but it can be measured only as it reflects light or alters the light of distant stars. No one has ever seen it, and what it is made of is still a matter of speculation. Some say it's made of metals, graphite, or silicates, while others argue that it's more like ice. All I know is that cosmic dust is one of the few dusts I don't have to apologize for when friends come knocking at my door.

EPILOGUE ON THE HUMAN HOUSE

At least once every year, I move out of my comfortable old farm-house and camp in the woods across my brook. The objective of this brief retreat from domestic convenience is to remind myself of what it feels like to live outdoors. Each year I find myself wondering about a different aspect of the indoor life I've left behind and the outdoor life I am leading. The year I was writing this book I thought about houses in general—why human beings live in them in the first place and how the houses we live in came to be the way they are.

Basically, a house is designed to shelter us from extremes of cold, hot, wet, or windy weather and to protect us from threats ranging in magnitude from predatory animals to biting or stinging insects. To serve these functions, a house can be anything from a hut to a palace, depending on the climate, available building materials, cultural beliefs, aesthetic convictions, socio-economic class, and architectural fashion.

It is customary to think of the ancestral human dwelling as a cave—and half a million years ago some human beings did indeed live in caves—but in areas where caves didn't exist, or in seasons and circumstances where caves were too confining or too stationary, primitive human beings also constructed windbreaks, lean-tos, and huts. Of these primitive structures, huts—being enclosed—are most like the houses we live in today.

The earliest huts were probably dome-shaped. Saplings or flexible branches were stuck into the ground in a circle and bent so that their tips could be tied together. This dome-shaped skeleton was then covered with grass, bark, leaves, or animal skins, depending on what was most available and most weatherproof.

Another type of primitive hut involved sinking both ends of flexible saplings or branches into the ground to create a series of parallel arches. These elongated, semicylindrical huts looked like miniature versions of the quonset huts that the U. S. military introduced during World War II.

Interestingly enough, the European settlers who came to this continent during the 1600s encountered both of these primitive hut types in northeastern North America. The Algonquin Indians, who lived in what has become Massachusetts, built dome-shaped wigwams, and the Iroquois in what has become New York State built semicylindrical long houses. But European architecture had evolved beyond such primitive shelters, and European settlers wanted semblances of their old homes in their new country.

The settlers came from many different architectural traditions, but timber and bricks had long since replaced the circles and arcs of primitive huts with the squares and rectangles of modern buildings. The preferred floor plan of colonial homes was rectangular because long side walls and short end walls provided the most efficient way to support a steeply pitched, weather-shedding roof.

In New England, timber-framed houses sided with clapboards and roofed with shingles predominated. Wood was the construction material of choice because there wasn't enough lime in the forested interior to make good mortar for laying bricks, and, furthermore, some of the early New Englanders believed bricks created an unhealthy environment to live in. Settlers farther south had no such reservations and, in Virginia, built themselves substantial brick houses to withstand the moister climate.

It took a group of Swedes who settled on Delaware Bay in 1638 to introduce the log cabin, which became the favored dwelling of the American pioneer, except in the prairies, where there were no trees. In the prairies, the sod house evolved, replacing the conical, wind-resistant, and easily transported teepees of the buffalo-following natives who preceded the European settlers.

In thinking about the many different forms that shelters have taken, I realized that the problem with modern houses and modern buildings in general is that they have evolved so far beyond merely sheltering us from undesirable weather, predators, and insects that they tempt us to stay indoors too much of the time.

What I learned from living outdoors for two weeks during the time I was writing this book is that houses, once they evolved beyond simple shelters, have actually done us a disservice. Anything that separates us unnecessarily from the outdoors both traps us among our own constructs—intellectual, physical, and ecological—and also deprives us of the contact we need with the integrating power of the natural world.

APPENDIX I: CHRONOLOGY

12,000 B.C.	Dog domesticated
5000 B.C.	Cotton grown and spun in Mexico
3000 B.C.	Cotton grown and spun in Pakistan
	Guinea pig domesticated in Peru
2500 B.C.	Cotton grown and spun in Peru
2000 B.C.	Cat domesticated in Egypt
A.D. 960–1279	Goldfish bred in ponds in China (Sung Dynasty)
1330	Goldfish kept in bowls in Peking
Early 1500s	Caged canaries sold in Europe
1532	Pizarro conquered the Incas
1564	Deposit of almost pure graphite discovered in Borrowdale, England
1565	Lead pencil first described
Around 1580	Guinea pig arrived in Europe
1638	Swedes built log cabins on Delaware Bay

1658	Broomcorn first described in Italy
1662	Royal Society of London chartered—Robert Hooke appointed Curator of Experiments
1665	Robert Hooke's *Micrographia* published—the word *cell* used to describe what cork looked like under his microscope
1680	Under *his* microscope, Anton van Leeuwenhoek saw "animalcules" in fermenting beer
1682	Claim that cedar would protect woolens against insect damage published
1691	Goldfish reached Great Britain
1725	Benjamin Franklin plucked a seed from the whisk broom of an English lady friend and planted the first broomcorn in North America
1730s	Rubber arrived in Europe (from South America)
1770	Joseph Priestley, in the first edition of *A Familiar Introduction to the Theory and Practice of Perspective,* recommended using rubber to erase pencil lines
1779	Swedish chemist Karl Wilhelm Scheele proved that the "lead" in lead pencils was actually carbon
1780	Guinea pig first used in laboratory
1789	Mineral that had been used to make pencils for over two centuries officially named graphite
1793	Heart-leaved philodendron brought to England by Captain Bligh
1795	Napoleon commissioned Nicholas Jacques Conte to develop a French pencil
1823	Charles Macintosh created the first rubber raincoat
1828	Goethe received a spider plant and distributed spiderlings to his friends
1837	Henry David Thoreau graduated from Harvard and joined his father in pencil making
1840	Parakeet brought to Europe (from Australia)
1844	Charles Goodyear patented the vulcanization of rubber
1857	Louis Pasteur first presented his theory of fermentation in "Sur la fermentation lactique" ("On Lactic Fermentation")
1858	Idea of attaching an eraser to a pencil patented
September 21, 1859	Thoreau's witch hazel seeds exploded indoors

1859	Edwin L. Drake struck oil in Titusville, Pennsylvania
1866	The Reverend Robert J. L. Guppy sent the first guppies to England
1868	Charles Darwin published *The Variation of Animals and Plants Under Domestication*; Canaries arrived in the United States
1876	Pasteur's *Études sur la Bière* proved once and for all that yeasts were alive; Henry Wickham brought 70,000 rubber seeds to England
1878	Goldfish arrived in the United States
1883	Krakatoa erupted, spewing 5 cubic miles (21 cubic km) of dust into the atmosphere
Early 1890s	Loofahs first grown for industrial use (in Japan)
1892	African violet discovered in East Africa
1894	Boston fern noticed among sword ferns shipped to Boston
1901	Gusher came in at Spindletop, Texas
1909	Thomas Hunt Morgan began his studies of fruit fly genetics
1930	Golden hamster discovered near Aleppo, Syria
1933	Morgan received Nobel Prize for his studies of fruit fly genetics
1936	Heart-leaved philodendron went on sale at five and dimes; Cellulose sponges became available
1938	Golden hamsters imported into the United States
1939	Blight destroyed sponges growing in the Caribbean
1949	Aloe vera first grown commercially in the Rio Grande Valley
1954	22 Mongolian gerbils arrived in the United States
1973	Oil crisis
1980s	1500 acres of aloe vera growing in the Rio Grande Valley

APPENDIX II: LATIN NAMES

An Indoor Animal Kingdom

Cat	*Felis catus*
Flea	*Ctenocephalides felis*
Dog	*Canis familiaris*
Guinea Pig	*Cavia porcellus*
Hamster	*Mesocricetus auratus*
Gerbil	*Meriones unguiculatus*
Canary	*Serinus canarius*
Parakeet	*Melopsittacus undulatus*
American Chameleon	*Anolis carolinensis*
Goldfish	*Carassius auratus*
Guppy	*Poecilia reticulata*
Sea Horse	*Hippocampus zosterae*

An Indoor Plant Kingdom

African Violet	*Saintpaulia ionantha*
Heart-Leaved Philodendron	*Philodendron scandens*
Wandering Jew	*Tradescantia fluminensis*
Jade Plant	*Crassula argentea*
Grape Ivy	*Cissus rhombifolia*
English Ivy	*Hedera helix*
Spider Plant	*Chlorophytum comosum*
Boston Fern	*Nephrolepis exaltata*
Narcissus	*Narcissus tazetta*
Amaryllis	*Hippeastrum vittatum*
Aloe Vera	*Aloe barbadensis*
Golden Barrel Cactus	*Echinocactus grusonii*

Other Kingdoms in the Kitchen

Yogurt	*Lactobacillus bulgaricus*
	Streptococcus thermophilus
Yeast	*Saccharomyces cerevisiae*
Bread Mold	*Rhizopus stolonifer*
Mildew	*Penicillium* spp.
Rotten Apples	*Alternaria* spp.
	Botrytis cinerea
	Cladosporium herbarum
	Cylindrocarpon mali and *Nectria galligena*
	Fusarium spp.
	Gloeosporium spp.
	Monilia fructigena and *Sclerotinia fructigena*
	Penicillium expansum and other spp.
	Phomopsis mali
	Trichothecium roseum

Household Ecology

Fruit Fly	*Drosophila melanogaster*
Housefly	*Musca domestica*
Ant	*Monomorium minimum*

Cockroach	*Blatella germanica*
House Mouse	*Mus musculus*
Deer Mouse	*Peromyscus maniculatus*
Polistes Wasp	*Polistes fuscatus*
Cluster Fly	*Pollenia rudis*
House Cricket	*Acheta domesticus*
Silverfish	*Lepisma saccharina*
Sowbug	*Oniscus asellus*
Centipede	*Scutigera coleoptrata*
House Spider	*Achaearanea tepidariorum*
Clothes Moth	*Tineola bisselliela*
Carpenter Ant	*Camponotus pennsylvanicus*
Termite	*Reticulitermes flavipes*
Bat	*Myotis lucifugus*
Chimney Swift	*Chaetura pelagica*

Household Natural History

Sponge	*Hippiospongia lachne*
Loofah	*Luffa aegyptiaca*
Cotton	*Gossypium hirsutum*
Witch Hazel	*Hamamelis virginiana*
Cedar	*Juniperus virginiana*
Cork	*Quercus suber*
Chopsticks (Bamboo)	*Arundinaria* spp.
	Bambusa spp.
	Dendrocalamus spp.
	Phyllostachys spp.
Broom (Broomcorn)	*Sorghum vulgare* var. technicum
Pencil	*Graphite*
Eraser (Rubber)	*Hevea brasiliensis*

SUGGESTED READING

An Indoor Animal Kingdom

Unfortunately, there is no comprehensive history of pets, but I did discover several interesting books and essays on domestication in my search for information on what I came to call "petification."

Charles Darwin's two-volume work, *The Variation of Animals and Plants Under Domestication* (New York: AMS Press, 1972), is a classic. He includes dogs, cats, canaries, goldfish, and guinea pigs among the hundreds of other animals he studied. Darwin's thoughts on domestication, variation, and selective breeding were closely related to his theory of natural selection. Basically, he tried to use what he could observe in domesticated life forms to understand what happens in nature—and he wasn't too far off according to modern geneticists. For an evaluation of Darwin's work, read Arne Müntzing's "Darwin's Views on Variation Under Domestication in Light of Present-Day Knowledge," *American Philosophical Society Proceedings*, 103 (1959), 190–220.

One of the works Darwin read while he was thinking about domestication still makes provocative reading today. In 1865, Francis Galton wrote an essay that proposed six conditions for successful domestication. In his own approximate words, the animal should be hardy, have an inborn liking for man, be comfort-loving, be found useful to the savages, breed freely, and be easy to tend. Galton also comments on the tendency of "savages" (his word) to keep pets, arguing that such primitive "petification" (my word) led to all the ancient domestications. Galton's essay, which originally appeared in the *Transactions of the Ethnological Society*, has been reprinted in *Inquiries into Human Faculty and Its Development* (New York: AMS Press, 1973).

Friedrich Eberhard Zeuner's *A History of Domesticated Animals* (London: Hutchinson, 1963) offers authoritative information based on more recent studies. Toward the end of his book, Zeuner talks about the human species' self-domestication—a concept I find intriguing. By "self-domestication" Zeuner means "that remarkable condition of some members of a species living at the expense of others, the whole forming a social unit in which all benefit from the division of labour." He compares human beings to honeybees in this respect, commenting that in bees "the process of self-domestication has gone much further than in man."

Each of our modern pets comes with its own legends and literature, available in the form of pamphlets, handbooks, encyclopedias, children's books, and fond biographies of beloved pets. T.F.H. Publications of Neptune, New Jersey, seems to have a book or booklet on the care and feeding of just about any pet you might be interested in. These inexpensive publications are available at pet stores. The reference section of your local library is another good place to look for both general and specialized books on pets and pet care.

If you want to read just one good book on many different pets, the best choice is Frances N. Chrystie's *Pets: A Complete Handbook on the Care, Understanding, and Appreciation of All Kinds of Pets*, Revised 3rd Edition (Boston: Little, Brown and Company, 1974). This book is addressed to young readers, but every adult who is considering buying a pet for a child should read it cover to cover. Chrystie emphasizes being realistic in the choice of a pet and responsible in its care.

Two other books I encountered in my reading seem worth mentioning. Thomas E. Gaddis' *Birdman of Alcatraz* (Mattituck, N.Y.: Aeonian Press, 1955) introduced me to the whole field of pets in therapy and rehabilitation. I found the story of Robert Stroud and his canaries more moving than the psychology books I read. Finally, Lee Edwards Benning's *The Pet Profiteers: The Exploitation of Pet Owners and Pets in America* (New York: Quadrangle/New York Times Books, 1976) is sobering. It touches on every abuse our pet-loving culture has managed to come up with—puppy mills, dog fights, and petnapping, to name just a few. The author's declared objective is to put America's human "pet parasites" out of business.

An Indoor Plant Kingdom

As with pets, I had difficulty finding a comprehensive history of houseplants. I was almost ready to give up the search, when a book that had been published in England was published in this country—and hence became available to me. Anthony Huxley's *The World Guide to House Plants* (New York: Charles Scribner's Sons, 1983) offered me exactly the kind of information I was looking for. Unfortunately, Huxley does not give an equal amount of information for every plant I had chosen to write about, so I am still haunted by a few houseplant mysteries. Another book that covers more plants but says less about each is Alfred Byrd Graf's *Exotic Plant Manual* (East Rutherford, N.J.: Roehrs Company, 1978).

Two other books offer official information on cultivated plants—such as Latin names and where the ancestral plants were discovered—and good quality black-and-white illustrations. *Hortus Third: A Concise Dictionary of Plants Cultivated in the United States and Canada* (New York: Macmillan Publishing Company, 1976) is a basic reference. Its forerunner, *The Standard Cyclopedia of Horticulture* by Liberty Hyde Bailey (New York: The Macmillan Company, 1914–1917), is of historic and aesthetic interest. Its six volumes offer many more illustrations.

For how to take care of houseplants, the best and most comprehensive guide is *Rodale's Encyclopedia of Indoor Gardening*, Anne M. Halpin, Editor (Emmaus, Pa.: Rodale Press, 1980). It includes a chapter on every subject you need to know about to take proper care of plants indoors, and it also lists individual houseplants with specific information on what each needs.

James Underwood Crockett's *Crockett's Indoor Garden* offers similar information in a different format. His book is arranged by months of the year, with how-to information and discussions of individual plants buried in the text. If you're looking for monthly houseplant activities, this book will keep you busy throughout the year, but if you're looking for information on what to do with a specific plant, you'll have to use the index to find the relevant passage.

Yet another approach to houseplants is the north-south-east-west approach of *House Plants for Five Exposures* by George Taloumis (New York: Abelard Schuman, 1973). Taloumis explains which plants need which kind of light, the fifth exposure being what he calls the "decorative exposure"—a table, shelf, or plant stand a distance from any particular window.

If you're worried about the health of your houseplants, Jean Blashfield's *The Healthy House Plant: A Guide to the Prevention, Detection, and Cure of Pests and Diseases* (Boston: Little, Brown and Company, 1980) will be helpful. And if you're into biological control of the insects who are most likely to attack your houseplants, you'll want to read *Windowsill Ecology* by William H. Jordan, Jr. (Emmaus, Pa.: Rodale Press, 1977).

Other Kingdoms in the Kitchen

On Food and Cooking: The Science and Lore of the Kitchen by Harold McGee (New York: Charles Scribner's Sons, 1984) offers detailed explanations—and amazing photographs—of various food-related microorganisms, such as the bacteria that turn milk into yogurt and the yeasts that turn grape juice into wine. A similar but shorter and easier-to-read book is *Science Experiments You Can Eat* by Vicki Cobb (New York: J.B. Lippincott/ A Harper Trophy Book, 1972). This clearly written children's book is full of edible activities using ordinary kitchen equipment and readily available ingredients.

While I was researching microorganisms, I discovered the writings of three scientists who deserve more popular attention. In 1665, the Royal Society of London published Robert Hooke's delightful accounts of what he saw through his primitive microscope in a book called *Micrographia, or Some Physiological Descriptions of Minute Bodies Made by Magnifying Glasses with Observations and Inquiries Thereupon* (New York: Dover Publications Inc., 1961). In addition to

looking at a microorganism—a bit of mildew growing on leather—
Hooke looked at handwriting and printed punctuation marks, a thin
slice of cork, a sponge, flies, silverfish, and fleas.

In 1673, Anton van Leeuwenhoek, who lived in the Neth-
erlands, began writing long letters to the Royal Society—some of
them addressed directly to Robert Hooke—describing the various phenomena and organisms *he* was seeing through *his* homemade microscopes. He reported seeing "animalcules" in fermenting beer, and he
detailed the four-stage life cycle of the flea. So far, eleven volumes of
The Collected Letters of Antoni van Leeuwenhoek have been edited and
translated by a Committee of Dutch Scientists (Amsterdam: Swets
and Zeitlinger, Ltd., 1939–1983), and some of these letters are great
fun to read.

Two centuries after Hooke and Leeuwenhoek, Louis Pasteur
studied microorganisms more scientifically. In his 1857 essay "On
Lactic Fermentation" (reprinted in *Pasteur's Study of Fermentation*,
Harvard Case Studies in Experimental Science, James Bryant
Conant, ed., Cambridge: Harvard University Press, 1952), Pasteur
first explained his organic theory of fermentation, but it took him
twenty years to win over the scientific community. Finally, in a book-
length study of beer (*Études sur la Bière*, 1876, translated as *Studies on
Fermentation*, New York: Kraus Reprint Company, 1969), Pasteur
demonstrated once and for all that the organisms responsible for fermentation are alive—as Leeuwenhoek had originally contended—and
that fermentation results from their life processes.

Household Ecology

Karl von Frisch's *Twelve Little Housemates*, Enlarged and Revised Edition (Oxford: Pergamon Press, 1978) is a popular classic by
a German entomologist best known for his studies of honeybees. Although Frisch focuses on the arthropods who invade European houses,
much of what he says is relevant to our North American arthropods.
His housemates include houseflies, fleas, clothes moths, cockroaches,
ants, silverfish, and spiders.

John C. McLoughlin's *The Animals Among Us: Wildlife in
the City* (New York: The Viking Press, 1978) is mostly about animals
who live outdoors, but his chapters on mice and "The Lesser Ten-
ants" (i.e., arthropods) focus on the indoors. Helen Ross Russell's
City Critters, Revised Edition (Homer, N.Y.: American Nature Study
Society, 1975) is similar but goes into less detail.

Ephraim Porter Felt's *Household and Camp Insects* (Albany: The University of the State of New York, 1917) is of historic interest for its descriptions of World War I era pest controls, which were based on arsenic, lethal-sounding concoctions of benzine, napthalene, creosote, and formaldehyde, and fumigations involving sulphur and hydrocyanic acid. Despite all. the chemicals, this New York State Museum Bulletin is full of precise accounts of common household insects, including detailed information on their life cycles.

I'm not sure *archy & mehitabel* (Garden City, N.Y.: Doubleday and Company, 1930) belongs among these suggested readings, but I would feel remiss if I neglected to mention North America's most famous cockroach. Several of Don Marquis' books are still in print and are an excellent antidote to excessive seriousness.

Household Natural History

My research on natural products took me into the fascinating world of economic botany. The journal entitled *Economic Botany* offered substantial articles on several of the subjects I had chosen for my book.

Richard M. Klein's *The Green World: An Introduction to Plants and People* (New York: Harper and Row Publishers, 1979) helped me with both kitchen microorganisms (bread, beer, and wine) and household products (cotton and rubber). The conversational tone and attractive illustrations make this college-level textbook a pleasure to read.

Another textbook, somewhat drier but more comprehensive in content, is Albert Frederick Hill's *Economic Botany: A Textbook of Useful Plants and Plant Products*, Second Edition (New York: McGraw-Hill, 1952). Hill discusses just about every plant, plant part, and plant product the human species has learned to use.

The Indoors

The best book on the indoors as a special environment is a recent children's book, which, unfortunately, has already gone out of print. Linda Allison's *Wild Inside: Sierra Club's Guide to the Great Indoors* (San Francisco/New York: Sierra Club Books/Charles Scribner's Sons, 1979) is lively, humorous, challenging, and full of practical information, activities, and send-aways. The text covers not only biology—the plants and animals who live indoors—but also geology and physics. Although written for children, this book is an excellent resource for parents, teachers, or anyone else who works with young people.

Two books by entomologists offer different ways of looking at specific houses. Vincent G. Dethier's *The Ecology of a Summer House* (Amherst, Ma.: The University of Massachusetts Press, 1984) describes the residents of his own vacation house in East Bluehill, Maine. Dethier writes engagingly, and the information he provides is authoritative.

George Ordish's *The Living American House: The 350-Year Story of a House, An Ecological Study* (New York: William Morrow & Company, Inc., 1981) lacks the charm of Dethier's more personal study, but what it lacks in charm, it makes up for in history. As the times changed, so did the nonhuman inhabitants of the house Ordish describes, which still stands in Duxbury, Massachusetts. Ordish tries to enliven his ecological history by telling tales of the different people who lived in the house, but the real interest of this book is its analysis of the different animals.

Last but not least, Charles A. Monagan's amusing book, *The Reluctant Naturalist: An Unnatural Field Guide to the Natural World* (New York: Atheneum, 1984) includes a short chapter entitled "The Indoor Naturalist." Because Monagan depicts himself as horrified by nature in general, he gets a lot of mileage out of molds, houseflies, spiders, mice, ants, silverfish, cockroaches, moths—and monsters.

INDEX

Printed in the United States
1483600004B/265

9 780595 167555